AF784447

SULPHUR IN AUSTRALASIAN AGRICULTURE

SULPHUR IN AUSTRALASIAN AGRICULTURE

Edited by
K. D. McLachlan

SYDNEY UNIVERSITY PRESS

SYDNEY UNIVERSITY PRESS
Press Building, University of Sydney

UNITED KINGDOM, EUROPE, MIDDLE EAST, AFRICA, CARIBBEAN
Prentice/Hall International, International Book Distributors Ltd
Hemel Hempstead, England
NORTH AND SOUTH AMERICA
International Scholarly Book Services, Inc., Portland, Oregon

First published 1975
© Sydney University Press 1975
Library of Congress Catalog Card Number 73-94024
ISBN 0 424 06850 8

This book is supported by money from
THE ELEANOR SOPHIA WOOD BEQUEST

Printed in Australia by John Sands Pty Ltd, Halstead Press Division

CONTENTS

TABLES

FIGURES

PREFACE

This book, one of two issued in association with the symposium on *Sulphur in Australasian Agriculture* held in Canberra during February 1974, is a record of the papers presented at the meeting. In them, active workers in sulphur research have reviewed basic knowledge and recent findings relating to the role of sulphur in the agricultural enterprise. The second book, *Handbook on Sulphur in Australian Agriculture*, contains factual data relating to sulphur in the various regions of the Australian environment.

Although half the free world sulphur production finds its way into fertilizer each year, sulphur as a fertilizer has traditionally been portrayed in a secondary role. It only assumes importance in such times as shortage of supply, or when social changes, e.g., anti-pollution policies, alter the emission patterns of industrial complexes. One of the purposes of this book was to show sulphur in its rightful role as a significant essential element for agriculture. The role of the book is therefore an integrating one. It brings together, within one cover and in review form, a wide range of topics affecting sulphur in agricultural production. It reviews the basic principles affecting the world at large, but in doing so places some emphasis on the peculiarities of the Australasian environment, thereby giving the book a character of its own.

The volume traces the story of sulphur from its origins and availability as a resource for the fertilizer industry, through its transformations in the biosphere, to its use by plants and animals for the ultimate benefit of man. It is the result of the efforts of scientists, technologists and practising agriculturists who were asked to cover particular aspects that lay within the realm of their expertise. A complete documentation was the aim.

The volume also deals with the more practical aspects of sulphur in agriculture, e.g., its role in ameliorating soil physical conditions, its role in restoring soil fertility, methods of assay for sulphur deficiency, and an evaluation of the relative merits of plant and soil tests. It looks at the sulphur sources available to industry, new developments in fertilizer manufacture, the problems associated with the development of sulphur fertilizers, and goes as far as suggesting the types of sulphur fertilizers that ought to be made to satisfy the needs of the Australian environment.

The interchange of ideas between participants was a significant feature of this inter-disciplinary symposium. While these discussions are not recorded here, the general tenor of the articles, both basic and applied, is such that this book should prove of value to the student, researcher, industrialist, and practising agriculturist alike. For all will find within its pages, material directly bearing on their own particular interests.

It gives me great pleasure to be able to express appreciation of the generous financial support given to the Symposium by:

Adelaide & Wallaroo Fertilizers Ltd
Australian Fertilizers Ltd
The Broken Hill Proprietary Coy Ltd
B.H. South Ltd
The Commonwealth Development Bank of Australia
CSBP and Farmers Ltd
CSIRO Division of Plant Industry
Electrolytic Zinc Company of Australasia Ltd
The National Science Foundation, Washington, D.C., U.S.A.
The Phosphate Cooperative Company of Australia Ltd, and
The Sulphur Institute, Washington D.C., U.S.A.

The assistance of these commercial enterprises and government institutions turned the seed of an idea sown by The Sulphur Institute into a viable entity, and brought it to fruition in a highly effective way.

Finally, I wish to record my thanks to the several committees responsible for the smooth operation of the Symposium, but in doing so it would be remiss of me not to pay particular tribute to J. R. Freney, J. S. Glendinning, W. Hartley, K. Spencer, S. L. Tisdale and C. H. Williams all of whom have been untiring in their efforts to ensure its success.

February 1974 K. D. McLachlan

CONTRIBUTORS

Anderson, J. W., Botany Department, LaTrobe University, Bundoora, Victoria
Andrew, C. S., CSIRO, Division of Tropical Agronomy, St Lucia, Queensland
Barrow, N. J., CSIRO, Division of Land Resources Management, Wembley, Western Australia
Bell, G. K., Australian Fertilizers Ltd, Pt Kembla, New South Wales
Bixby, D. W., The Sulphur Institute, Washington, D.C., United States of America
Blair, G. J., Agronomy Department, University of New England, Armidale, New South Wales
Bouma, D., CSIRO, Division of Plant Industry, Canberra City, Australian Capital Territory
Cooper, D. M., Ruakura Agricultural Research Centre, Hamilton, New Zealand
Crack, B. J., Department of Primary Industries, Indooroopilly, Queensland
Downes, A. M., CSIRO, Division of Animal Physiology, Prospect, New South Wales
Evans, L. T., CSIRO, Division of Plant Industry, Canberra City, Australian Capital Territory
Freney, J. R., CSIRO, Division of Plant Industry, Canberra City, Australian Capital Territory
Gregg, P. E. H., Ministry of Agriculture and Fisheries, New Zealand
Humphreys, F. R., Forestry Commission of New South Wales, Epping, New South Wales
Johnson, A., The Naroo Pastoral Co. Pty Ltd, G.P.O. Box 3732, Sydney, New South Wales
Jones, M. B., University of California, Hopland Field Station, Hopland, California, United States of America
Jones, R. K., CSIRO, Division of Tropical Agronomy, Townsville, Queensland
Kelly, J., Forestry Commission of New South Wales, Epping, New South Wales
Kilmer, V. J., National Fertilizer Development Center, TVA, Muscle Shoals, Alabama, United States of America
Lambert, Marcia J., Forestry Commission of New South Wales, Epping, New South Wales
Langlands, J. P., CSIRO, Division of Animal Physiology, Armidale, New South Wales
Loveday, J., CSIRO, Division of Soils, Canberra City, Australian Capital Territory
McLachlan, K. D., CSIRO, Division of Plant Industry, Canberra City, Australian Capital Territory
McSweeney, G., New Zealand Fertilizer Manufacturers' Research Association, Papatoetoe, New Zealand
Moir, R. J., Department of Animal Science and Production, University of Western Australia, Nedlands, Western Australia
Muller, F. B., New Zealand Fertilizer Manufacturers' Research Association, Papatoetoe, New Zealand

Nicolson, A. J., CSIRO, Division of Tropical Agronomy, Mareeba, Queensland
Probert, M. E., CSIRO, Division of Soils, Townsville, Queensland
Reis, P. J., CSIRO, Division of Animal Physiology, Prospect, New South Wales
Reisenauer, H. M., Department of Soils and Plant Nutrition, University of California, Davis, California, United States of America
Rogers, J., New Zealand Fertilizer Manufacturers' Research Association, Papatoetoe, New Zealand
Saunders, W. M. H., Ruakura Agricultural Research Centre, Hamilton, New Zealand
Spencer, K., CSIRO, Division of Plant Industry, Canberra City, Australian Capital Territory
Swaby, R. J., CSIRO, Division of Soils, Glen Osmond, South Australia
Till, A. R., CSIRO, Division of Animal Physiology, Armidale, New South Wales
Tisdale, S. L., The Sulphur Institute, Washington D.C., United States of America
Trudinger, P. A., The Baas Becking Geobiological Laboratory, Canberra City, Australian Capital Territory
Walker, T. W., Soil Science Department, Lincoln College, Canterbury, New Zealand
Weir, R. G., Department of Agriculture, Rydalmere, New South Wales
Williams, C. H., CSIRO, Division of Plant Industry, Canberra City, Australian Capital Territory.

1 Introduction

Sulphur in Agriculture

L. T. EVANS

One of the advantages of a symposium topic like sulphur is that it brings together research workers from many disciplines, as evident in previous symposia on sulphur at Muscle Shoals, Corvallis, Wexford and Versailles. The strong S-S bonds between these symposia have put conformational constraints on their structure, and on the introductory papers, as they do on enzymes. Nevertheless, each has reflected the particular interests of its organizers, and this Canberra symposium is no exception, concentrating as it does on soil and plant aspects, and on responses to sulphur in Australia and New Zealand.

The essentiality of sulphur as a plant nutrient has been known for more than 100 years, and Liebig (20) was aware of the close relation between nitrogen and sulphur in many plants. Early responses to gypsum, like Benjamin Franklin's lettering of a hillside with 'This Land Has Been Plastered', could have been due to either the S or the Ca applied. In 1899 Bogdanoff in Russia apparently reported responses by mustard and other crops to the application of Na_2SO_4, and increases in yield clearly attributable to the application of S were found in 1905 with maize and red clover grown on sandy soils with the plants protected from accessions of sulphur in rain (10). Since that time field responses to S fertilization have become increasingly widespread, especially in North America, Africa and Australasia.

The year 1929 saw responses to sulphur recorded in New Zealand by Doak, and by Tennent and Duff, and in Australia by Mary Glynne. In this latter case, however, the response of wheat to elemental S was probably due to a correction of manganese deficiency via an effect on soil pH and it was not until 1950 that sulphur deficiency was clearly demonstrated in Australia (1). By 1952 other members of the CSIRO Division of Plant Industry had found field responses to S on the Tablelands of N.S.W., in Queensland and in Western Australia, and so began the long programme of research on sulphur in the Division, a sponsor of this symposium.

Several reasons may be advanced for the increasing recognition of sulphur deficiency. Many cereals and other crops remove more than 20 kg S ha^{-1} each year, while for some Cruciferae and Liliaceae this loss may be as much as 70 kg ha^{-1} (3), far more than the usual annual accession of sulphur. In the U.S.A. crops remove about 2.3×10^6 tons S annually, yet only 1.3×10^6 tons S were applied in fertilizers in 1972. Accession of sulphur in rain on the 1.8×10^8 ha of cropland, in the U.S.A. may add a further 2×10^6 tons yr^{-1}, but leaching losses could equal this. Meat and wool production also entail losses of S. Secondly, recent increases in the use of high-analysis N and P fertilizers, to reduce freight and handling costs, have resulted in a substantial reduction in the amount of sulphur applied in fertilizers. In the U.S.A., for example, this has decreased from 2.4×10^6 tons S in 1950 to only half that amount by 1972. The amount of S applied to crops as fungicides and insecticides has also declined. Thirdly, increased applications of other nutrients in fertilizers, and greater emphasis on high protein crops, both generate greater requirements for sulphur. In Australia, the use of superphosphate to increase clover growth and the nitrogen status of pastures has resulted in increased requirements for sulphur.

ATMOSPHERIC SULPHUR

The natural supply of S to the atmosphere, according to Robinson and Robbins (25), is 142×10^6 tons per year, of which about half comes from the sea, 44×10^6 tons as sulphate in spray and 30×10^6 tons as H_2S from oceanic decay. The H_2S is rapidly oxidized by atmospheric ozone. The terrestrial origin of the remainder is indicated by the fact that sulphate in rain has a much lower proportion of S^{34} than occurs in seawater, presumably due to extensive emission of H_2S from swamps and low-oxygen muds.

Anthropogenic emissions of S into the atmosphere, mainly SO_2, are now equivalent to about half the total natural emissions (17). The burning of coal, with an average S content of 2 per cent, is the main source, but about 20 per cent comes from petrol and 10 per cent from metal smelting. Although the anthropogenic component has increased rapidly in recent years, the greater use of low S-coals and more efficient stack recovery of S in future may reduce man's contribution of S to the atmosphere, while increasing drainage of swamps may also reduce the natural input.

To get the annual atmospheric contribution of 2.2×10^8 tons S yr^{-1} into perspective, we might compare it with the amount of S incorporated by plants in association with photosynthesis. Rodin estimates that global photosynthesis produces 2.2-2.4×10^{11} tons dry matter per year. Allowing a sulphur content of 0.2 per cent, this would amount to 4.6×10^8 tons S per year. Atmospheric S could therefore contribute at least half of the total incorporation each year and, taking the rapid turnover of much of this into consideration, probably most of it. Some years ago Eriksson suggested that S derived from rock weathering could account for only a small proportion of the S in rivers, most of it coming from the atmosphere (8).

Thus, the atmospheric cycle is almost as predominant for sulphur as it is for nitrogen, and far more so than for other nutrients. Moreover, plants can take up S directly from the atmosphere, without the intervention of soil. Thomas and Hill (27) showed that lucerne could take up a substantial proportion of its S as SO_2 from the atmosphere without injury, and this finding has subsequently been confirmed for other crop plants such as cotton (22) and rape (4). Uptake of atmospheric SO_2 is able to compensate for soil S deficiency (6). Indeed, SO_2 and H_2S can be used by sunflowers as their only source of sulphur without affecting their normal growth (12).

Although bush fires may generate considerable SO_2 in the short term, atmospheric S levels are probably low in most Australian agricultural areas, reflecting their low rainfall and considerable distance from both sea and industry. More heavily industrialized countries are well aware of the potential fertilizer value of sulphur from the atmosphere. There is a considerable margin between SO_2 concentrations which promote plant growth and those at which injury occurs. Aerial distribution of S should therefore be feasible, but foreseeable atmospheric levels in most agricultural areas will be insufficient to sustain highly productive crops and pastures (24).

SOIL SULPHUR

In many soils organic sulphur is the predominant component, the soluble and adsorbed sulphate available for uptake by plants constituting only a small proportion of the total soil S, often less than 10 per cent. Sulphate levels are often subject to considerable seasonal fluctuation, and may fall to very low values, for example to less than 1 ppm in the sandy soils of Western Australia during the wet winter when mineralization is minimal and leaching losses are pronounced.

The organic S of soils is derived from the decomposition of plant and microbial matter, and from the excreta of grazing animals. Where the sulphur content is low in plant materials, less than say 0.15 per cent, it may limit the rate at which decomposition can occur.

Soil organic matter is enriched in sulphur relative to nitrogen, with an N:S ratio of about 8 compared with a ratio of 15 to 17 in most plant proteins, and of more than 10 even in plant material abundantly supplied with sulphur (5). The relative enrichment in S implies that organic N must often mineralize faster than organic S in the early stages of plant decomposition (2,14). Presumably, the proteins which comprise most of the S in plants and soil micro-organisms are more rapidly broken down than other S-containing compounds in them, the amino acids being recycled. This is suggested by the fact that Raney nickel-reducible S, mainly cystine and methionine, often represents only 20 to 30 per cent of the organic S in soils (cf. Williams, p. 24). The remainder is divided about equally between hydriodic acid-reducible compounds (such as ester sulphates) and carbon-bonded S compounds not reducible by HI. Some of these latter compounds may derive fairly directly from S-containing components of plants and micro-organisms, but others may be heterogeneous, complex and resistant products of multiple soil transformations. The HI-reducible S, which may comprise more than 60 per cent of the total organic S in some soils, may include compounds such as sulphated polysaccharides, choline sulphate and lignin bisulphites (13).

Improved fractionation procedures and a greater knowledge of what happens to specific S-compounds in soils are clearly pre-requisite to a better understanding of the nature, role and dynamics of organic S in soils.

SULPHUR IN PLANTS

Nearly all the S taken up by roots is as sulphate except in plants like rice growing on flooded soils, which take up more reduced forms. Only a small proportion of the sulphate is reduced in the roots, less than 0.4 per cent according to Pate (23), and probably far less than the proportion of nitrate reduced in the roots. Instead, the sulphate is translocated to the leaves and reduced in the chloroplasts, with reduced ferredoxin and $NADPH_2$ as electron donors. Possibly because they are the main site of sulphate reduction, the chloroplasts have a well-known advantage over the cytoplasm in the retention of their S (29), Fraction I protein being considerably richer in cystine and methionine than leaf protein as a whole (18).

Except at high levels of S nutrition in some Cruciferous and Liliaceous plants, most plant S occurs in proteins. Cysteine plays a major role in determining, through S-S bonds, the three dimensional structure of proteins, while the sulfhydryl groups of both methionine and cystine are important sites for metal cation binding. Consequently, the sulphur-containing amino acids play a key role in determining the activity of most enzymes. Their effects on protein structure and reactivity are also important in hardiness and resistance to stress, and possibly also in morphogenesis. Many other sulphur compounds are also crucial to metabolism, e.g., biotin, thiamine, glutathione, lipoic acid, co-enzyme A and ferredoxin. Others which may be even more abundant, but whose role is less clear, include the glucosinolates in many Cruciferae and the S-alkylcysteine sulphoxides in Liliaceae such as onions and garlic.

High levels of available S in soils do not seem to influence the proportion of high S proteins in plant tissue. In the leaves and roots of many species, for example, protein S increases in proportion to protein N over a wide range (9,26), implying that there is no increase in the proportion of high S proteins following heavy applications of S fertilizer. Changes in the relative proportions of the various seed proteins may be more likely, but have yet to be shown. Total N:total S ratios may vary widely, e.g., from 4 to 55 in the experiments of Stewart and Porter (26). At low available soil S levels, virtually all the plant S was in protein, whereas less than a quarter of the total N was in protein, there being a major accumulation of amides and amino acids. At high soil S levels, on the other hand, sulphate may reach high levels in plant tissues, as may the S-containing amino acids. In rape, for example, cystine accumulates at high S levels, but much of the methionine is converted to glucosinolates, especially at low N levels (16). In fact, the amount of methionine converted to glucosinolates in such plants exceeded that incorporated into leaf proteins.

Plant species differ considerably not only in the concentration of accessory organic sulphur compounds in them, but also in the tissue sulphate concentration they maintain at a particular level of sulphate in the nutrient solution (11). Cotton and tomatoes, for example, are heavy sulphate accumulators. However, some of this sulphate may be sequestered or compartmentalized to become unavailable. Randall (unpublished) has found the high contents of sulphate in old leaves of rape to remain there even when young leaves were developing under severe deficiency conditions.

As a result of these differences in the level of sulphate and organic sulphur compounds, species differ considerably in their critical percentage S, from less than 0.1 per cent in some cereals to more than 0.7 per cent in rape and kale. Likewise, the requirement for S in solution culture ranges from less than 1 ppm for wheat to 4 ppm for several clovers and 8 ppm for rape and lucerne (p. 102). Low S levels may, consequently, favour grasses at the expense of clovers (28).

Thus, although sulphur plays many indispensable roles in the metabolism of all plants, species nevertheless differ greatly in their requirements for sulphur and in their ability to obtain it from their environment. We need to understand the physiological basis of these differences before we can meaningfully compare critical S and sulphate levels in different plants.

SULPHUR IN ANIMALS

Unlike plants, animals are unable to reduce, and therefore use sulphate, but in ruminants this is accomplished for them by their ruminal micro-organisms. However, ruminants tend to squander this advantage and do not recycle nearly as high a proportion of their S as of their N (p. 112), much of the intake of S being returned to the soil in urine.

Wool proteins contain 2.7 to 4.2 per cent (associated with 10 to 15 per cent cystine), and wool growth may often be limited by the supply of cystine and methionine (p. 119). Abomasal or subcutaneous supply of these amino acids can increase wool growth, but uptake of additional cystine and methionine via the rumen may merely result in their increased microbial degradation there. It is a challenging question whether plant characteristics could be selected or modified to provide more effective protection of these amino acids in their passage through the rumen.

There are, however, several adverse effects of high sulphur levels. High sulphide levels in the rumen may lead to feed refusal and respiratory distress (p. 121), and high sulphate levels may aggravate the adverse effects of molybdenum on copper retention. High sulphur and glucosinolate levels in brassicas are goitrogenic. Of considerable interest both in the southern tablelands of N.S.W. and elsewhere are the possible effects of raised soil sulphur levels on selenium uptake by plants and on the incidence of white muscle disease in ruminants. Sulphate and selenate are competitive for uptake by both plants and animals. Heavy sulphur applications were originally used to reduce selenium uptake by plants in areas where it occurred in toxic concentrations (15), but surface applications of S cannot, of course, reduce the uptake of Se at depth by plants. However, improved S levels may lead to an increase in clovers, which take up less selenium than grasses (7) and this, rather than competition between S and Se for uptake by plants, may lead to an increased incidence of white muscle disease (cf. (21)). In ruminant nutrition, however, there is clear evidence that higher sulphate levels can depress selenium uptake (p. 114).

Consequently, if more S fertilizers are applied to pastures in areas where the Se level is marginal, they may cause an increased incidence of white muscle disease. To supply Se bullets to stock, to all young lambs for example, may be impractical, in which case consideration may need to be given to the problems of adding Se to all S fertilizers. This too is a challenging problem, to develop procedures which walk the tightrope between Se deficiency on the one hand and toxicity on the other.

Sulphur, the brimstone of past times, constitutes 10 to 15 per cent of the earth's core (19) but only 0.06 per cent of its crust and about 0.07 per cent of the earth's biomass (8). It is a nutrient of crucial significance to both plants and animals. Fertilizers account for about 40 per cent of the total S used in the world, mostly as sulphuric acid used to dissolve phosphate rocks. Several factors point to an increased requirement for sulphur fertilizers in the future, rather than to a continuation of the decline of recent years. Although one of the more difficult nutrients to handle analytically, sulphur and its biological cycle will demand even more attention in the future.

REFERENCES

(1) Anderson, A. J. and Spencer, D. (1950). *Aust. Sci. Res.*, Ser. B, **3**: 431.
(2) Barrow, N. J. (1961). *Aust. J. Agric. Res.*, **12**: 644.
(3) Beaton, J. D. (1966). *Soil Sci.*, **101**: 267.
(4) Bromfield, A. R. (1972). *J. Agric. Sci.*, **78**: 343.
(5) Cowling, D. W. and Jones, L. H. P. (1970). *Soil Sci.*, **110**: 346.
(6) Cowling, D. W., Jones, L. H. P. and Lockyer, D. R. (1973). *Nature* (London), **243**: 479.
(7) Davies, E. B. and Watkinson, J. H. (1966), *N.Z. J. Agr. Res.*, **9**: 641.
(8) Deevey, E. S. (1970). *Sci. Amer.*, **223**(3): 149.
(9) Dijkshoorn, W. and van Wijk, A. L. (1967). *Pl. Soil*, **26**: 129.
(10) Dymond, T. S., Hughes, F. and Jupe, C. W. C. (1905). *J. Agron. Sci.*, **1**: 217.
(11) Eaton, F. M. (1966). In Chapman, H. D. (ed.), *Diagnostic Criteria for Plants and Soils*. University of California, p. 444.
(12) Faller, N. (1972). *Zeitschr. Pflanzenernahr. Bodenk.*, **131**: 120.
(13) Freney, J. R. (1961). *Aust. J. Agric. Res.*, **12**: 424.
(14) Freney, J. R. and Spencer, K. (1960). *Aust. J. Agric. Res.*, **11**: 339.
(15) Hurd-Karrer, A. (1937). *J. Agr. Res.*, **54**: 601.
(16) Josefsson, E. (1970). *J. Sci. Fd. Agric.*, **21**: 98.
(17) Junge, C. (1972). *Int. Symp. on Sulphur in Agriculture*, Versailles, 1970. INRA, p. 235.
(18) Kawashima, N. and Wildman, S. G. (1970). *Ann. Rev. Pl. Physiol.*, **21**: 325.
(19) King, D. A. and Ahrens, T. J. (1973). *Nature Physical Science*, **243**(127): 82.
(20) Liebig, J. (1859). *Letters on Modern Agriculture*. Blyth, J. (ed.). Walton & Maberly, London.
(21) Mathias, M. M., Allaway, W. H., Hogue, D. E., Marion, M. V. and Gardner, R. W. (1965). *J. Nutr.*, **86**: 213.
(22) Olsen, R. A. (1957). *Soil Sci.*, **84**: 107.
(23) Pate, J. S. (1965). *Science*, **149**: 547.
(24) Prince, R. and Ross, F. F. (1972). *Water Air Soil Pollut.*, **1**: 286.
(25) Robinson, E. and Robbins, R. C. (1968). *Sources, Abundance and Fate of Gaseous Atmospheric Pollutants*. Prepared for API (Stanford Res. Inst.).
(26) Stewart, B. A. and Porter, L. K. (1969). *Agron. J.*, **61**: 267.
(27) Thomas, M. D. and Hill, G. R. (1935). *Plant Physiol.*, **10**: 291.
(28) Walker, T. W. and Adams, A. F. R. (1958). *Pl. Soil*, **9**: 353.
(29) Williams, R. F. (1955). *Ann. Rev. Pl. Physiol.*, **6**: 25.

2 Forms, Availability and Transformations of Sulphur in Soils

The Biogeochemistry of Sulphur

P. A. TRUDINGER

Sulphur is the thirteenth most abundant element in the Earth's crust (7). At roughly 0.05 per cent it falls into the category of a trace element but as in the case of many other trace elements, the terrestrial abundance of sulphur bears little witness to its importance in natural phenomena. Geochemically sulphur must be regarded as one of the more significant elements being a constituent of many minerals of great economic value. Moreover, the physico-chemical conditions in sediments, soils and other natural environments are markedly influenced by biological activity, and transformations of sulphur compounds therefore contribute, often extensively, to the control of chemical events taking place within such environments.

Sulphur also plays an important role in living organisms. It is a component of several essential amino acids and vitamins as well as of a variety of organic compounds whose physiological significance remains to be determined (13).

The most abundant compound of sulphur is sulphate. Ocean waters, on average, contain about 0.27 per cent sulphate. In other natural waters the sulphate content generally ranges from about 5×10^{-5} per cent to 5×10^{-3} per cent but may reach 6 per cent in highly saline lakes and sediments (18). Numerous solid mineral sulphates have been described, the most important of which are anhydrite ($CaSO_4$) and gypsum ($CaSO_4.2H_2O$) (7). Other forms of major geochemical interest are elemental sulphur and the metallic sulphides. These are widely, though sparingly distributed in the lithosphere but in localized areas they may be concentrated to a sufficient extent to enable economic exploitation. Minor forms of inorganic sulphur include gaseous sulphur oxides from volcanic exhalations and hydrogen sulphide of volcanic, hydrothermal and biological origin.

In many soils, particularly those of humid regions, a major portion of the total sulphur may be present in organic form. The chemical nature of the material is uncertain but a considerable fraction appears to be sulphate covalently bound to certain compounds in the soil organic matter (14). This will be the subject of a separate paper in this Symposium (15) and in the subsequent discussion the term 'organic sulphur' will refer primarily to the reduced sulphur present in proteins and other constituents of living cells.

THE BIOGEOCHEMICAL CYCLE

Estimates of the magnitudes of terrestrial reservoirs of sulphur (Table 2-1) suggest that about 50 per cent of crustal sulphur is present in relatively mobile reservoirs represented by the sea, evaporites and sediments. Organic

TABLE 2-1
Estimated sulphur reservoirs

Pool	S (metric ton)	Reference
Total crustal sulphur	1.8×10^{16}	(17)
Atmosphere	3.6×10^{6}	
Sea	1.3×10^{15}	
Fresh water	3×10^{9}	
Evaporites	5×10^{15}	
Sediments	2.7×10^{15}	
Metamorphic and igneous rocks	7×10^{15}	
Land plants	6×10^{8}	(10)
Marine plants	2.4×10^{7}	
Dead organic matter	5×10^{9}	

matter, in the form of living organisms or dead material, is a quantitatively insignificant reservoir of sulphur.

In common with most other elements, however, the various forms of sulphur undergo continual transformation and interconversion. By virtue of their metabolic activities all organisms play some part in these processes, generally by altering the oxidation state of the element. The more important microbial transformations of sulphur are listed in Table 2-2. These involve the interconversion of sulphate with the most reduced forms of the element, sulphur and sulphide. Inorganic compounds containing sulphur at intermediate oxidation states, for example thiosulphate and polythionates, are also produced by micro-organisms (34) but the geochemical role of these compounds is uncertain. Significant concentrations of thiosulphate, however, have been detected in euxinic basins (43).

TABLE 2-2
Major microbiological transformations of sulphur*

Organisms	Reactions	
Most micro-organisms	$SO_4^{2-} \longrightarrow$ CYSTEINE (+ other organic S—compounds)	1
Many heterotrophs	CYSTEINE $\longrightarrow H_2S$	2
Heterotrophs	CYSTEINE $\longrightarrow SO_4^{2-}$ (via organic pathway)	3
Beggiatoa	$H_2S \longrightarrow S^0$	4
Dissimilatory sulphate reducers (*Desulfovibrio, Desulfotomaculum*)	$SO_4^{2-} \longrightarrow H_2S$	5
Chemosynthetic sulphur autotrophs	$H_2S \longrightarrow SO_4^{2-}$	6
	$H_2S \longrightarrow S^0$	4
e.g. *Thiobacillus*	$S^0 \longrightarrow SO_4^{2-}$	7
Photosynthetic sulphur autotrophs	$H_2S \longrightarrow S^0$	4
e.g. *Chlorobium, Chromatium*	$H_2S \longrightarrow SO_4^{2-}$	6

* Numbers refer to steps in Figure 2-1.

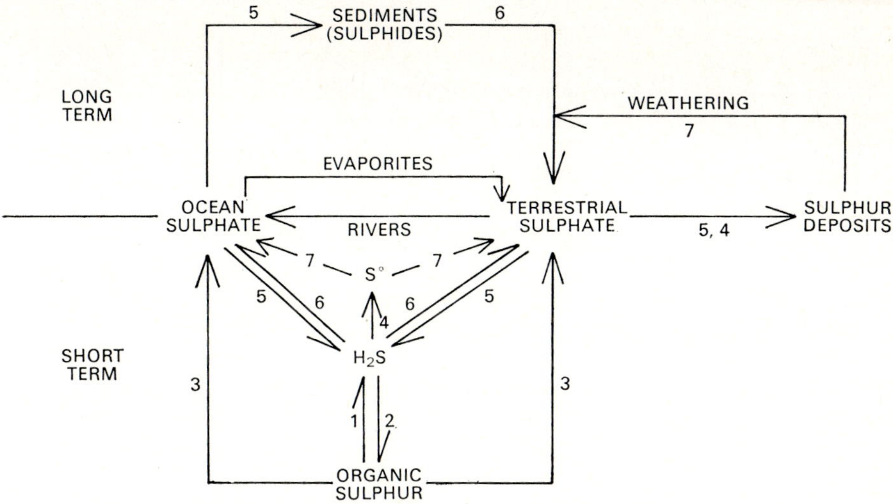

Figure 2-1 Simplified representation of the biogeochemical cycle of sulphur. Numbers refer to reactions in Table 2-2.

Figure 2-1 shows a simplified version of the geochemical cycle of sulphur and indicates those stages in the cycle in which micro-organisms are known or suspected to play a significant part. From the biogeochemical viewpoint the most important aspects of this cycle are: (1) the rapid interconversion of oxidized and reduced forms of sulphur which takes place in soils and in the waters and sediments of marine and fresh water environments; and (2) the long term cycle involving massive precipitation of sulphur and metallic sulphides in various geological periods and their subsequent oxidative weathering to sulphates.

Because it is common to virtually all micro-organisms and plants the reduction of sulphate to cysteine (assimilatory sulphate reduction) is probably a major biological process involved in the transformation of sulphur. Riley (32) estimated the total annual organic carbon production for the earth to be in the order of 1.5×10^{11} metric tons. Based on an average C:S ratio of 100 this would correspond to a rate of synthesis of organic sulphur of about $1\text{-}2 \times 10^9$ tons per annum.

The sulphur of fossil organic matter such as humic acids and kerogens may be partly derived from this source. Cysteine, or its oxidation product cystine, has also been shown to react with metallic ions under mild conditions to form insoluble metal sulphides (25). The major part of organic sulphur, however, is rapidly reconverted to sulphate by biochemical processes (see Table 2-2) and is unlikely to contribute significantly to the formation of extensive deposits of sulphur minerals. Eriksson (10,11) in studies on the world-wide yearly circulation of sulphur concluded that all of the organic sulphur synthesized by land plants is eventually remineralized and that only about 7 per cent of that produced by marine plants is incorporated into sediments. Consideration of specific environmental situations leads to similar conclusions. In a study of the carbon budget of the Black Sea, for example, Deuser (8) concluded that only about 4 per cent of the primary organic carbon resulting from photosynthetic carbon dioxide fixation was incorporated into the sediments, the remainder being metabolized in the overlying waters. We may assume that the extent of remineralization of organic sulphur is of a similar order of magnitude. Based on an annual rate of organic matter production of 100 g m^{-2} yr^{-1} (41) we may calculate, therefore, that the rate of incorporation of organic sulphur into the Black Sea sediments would be about 40 mg S m^{-2} yr^{-1}. The minimum rate of sedimentation in the euxinic zone of the Black Sea is in the order of 0.01 cm yr^{-1} (33) and, assuming a density of 1.0 for fresh sediment, we could not, therefore, expect the concentration of organic sulphur to exceed about 0.04 per cent.

More by intuition than from actual knowledge we suspect that the main biological processes affecting interchange between the principal sulphur reservoirs are: (1) the dissimilatory reduction of sulphate to hydrogen sulphide catalyzed by the anaerobic bacteria, *Desulfovibrio* and *Desulfotomaculum* (31); and (2) the lithotrophic oxidation of sulphide and elemental sulphur by the aerobic chemosynthetic bacteria (e.g. *Thiobacillus*) and the anaerobic photosynthetic bacteria (e.g. *Chromatium* and *Chlorobium*) (34). Several euxinic basins are remarkable examples of environments in which short-term cycling of sulphur has been established through the co-operative action of sulphate-reducing and sulphide-oxidizing bacteria. Figure 2-2, which is based on Sorokin's data for the Black Sea (40,42) and a number of fresh

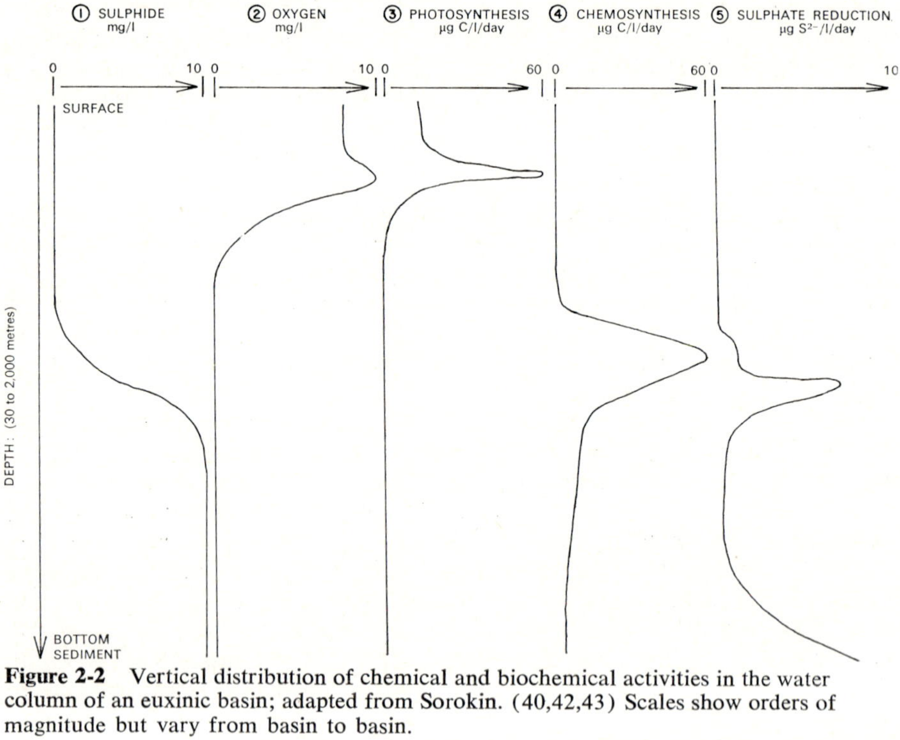

Figure 2-2 Vertical distribution of chemical and biochemical activities in the water column of an euxinic basin; adapted from Sorokin. (40,42,43) Scales show orders of magnitude but vary from basin to basin.

water lakes (43), shows the general distribution of some of the chemical and biochemical activities in the water columns of such basins. In the upper layers the waters are oxygenated (trace 2) and organic matter is produced from carbon dioxide and other nutrients by planktonic algae (trace 3). Below the photosynthetic zone organic matter is oxidized by a mixed heterotrophic microflora and as a result dissolved oxygen is rapidly depleted. Eventually, reducing conditions are established and there develops a zone of active sulphate reduction (trace 5) which is supported by the residue of dissolved organic matter. A second zone of sulphate reduction is found in the bottom sediments and at the sediment-water interface where sedimentation of organic particles and transport of fine upper silt layers from shallower parts of the basin produce a region rich in organic matter. A zone of chemosynthesis is established where hydrogen sulphide diffuses into the region of oxygen availability (trace 4). The nett result of these various processes is that a virtually steady state concentration of hydrogen sulphide is maintained in the lower anoxygenous waters of the basin (trace 1) (44).

DEPOSITION OF SULPHIDES

Biological activity is quite clearly the dominant factor in the short-term recycling of sulphur in environments such as euxinic basins, soils and sediments. It has also been estimated that in basins like the Black Sea the sulphur flux, about 20 g S m^{-2} yr^{-1} (40), is of the right order for the formation of a massive sulphide deposit if sufficient metal were available (48).

Iron sulphide deposition resulting from bacterial sulphate reduction is commonly observed in euxinic sediments, tidal flats and the littoral zones of oceans (37) while the 'biogenic' formation of sulphides of a number of base metals has been demonstrated experimentally (3,28). Palaeontogical evidence suggests that the depositional environments of many sedimentary sulphide ores, for example the McArthur River deposits in Northern Australia (4,29), supported an active biological community.

Observations such as those just described have led to the notion that the sulphide of some massive iron and base metal deposits may be of biological origin. At the present time, however, there is no unequivocal evidence for or against this view, a situation which arises largely from the lack of clear-cut criteria whereby biogenic sulphides, as distinct from those of chemical origin, may be recognized.

Recently the possible use of stable sulphur isotopes in determining the origin of sulphur minerals has gained prominence. Primordial terrestrial sulphur is thought to have had a constant isotopic composition close to that found in meteorites ($^{34}S/^{32}S = 0.045$). The isotopic composition of many present-day minerals differs widely from this value and it is considered that this is due to isotopic discrimination during transformation of sulphur from its primordial state. Variations are generally expressed in terms of $\delta^{34}S$ which relates the isotopic composition of sulphur minerals to a standard by the expression

$$\delta^{34}S(‰) = \left[\frac{^{34}S/^{32}S \text{ sample}}{^{34}S/^{32}S \text{ standard}} - 1 \right] \times 1{,}000$$

The isotopic composition of naturally occurring sulphur minerals is generally referred to that of the troilite in the Canon Diabolo meteorite. The $\delta^{34}S$ values for most sulphur minerals of igneous origin are relatively uniform and tend to average around -10 to $+10$‰ (45). On the other hand $\delta^{34}S$ values of sedimentary sulphides vary considerably both within and between deposits (roughly $+45$ to -50‰ overall) and on average tend towards negative values, that is towards enrichment in ^{32}S (45). A widely advocated explanation for this observation is that the sedimentary minerals were precipitated from hydrogen sulphide of biological origin.

Two basic mechanisms apply for isotopic discrimination during sulphate reduction. The first is isotope exchange

$$^{32}SO_4^{2-} + H_2{}^{34}S \rightleftharpoons {}^{34}SO_4^{2-} + H_2{}^{32}S$$

which, at equilibrium at 25°, would result in a $\delta^{34}S$ for H_2S of about -70‰ with respect to sulphate (49). This exchange reaction has not been realized in purely chemical systems but could, in theory, be catalyzed by enzymes of the dissimilatory sulphate reduction pathway, or by the co-operative action of sulphate-reducing and sulphide-oxidizing bacteria (47).

The second fractionation mechanism is kinetic and is probably due to differences in the strength of the S — O bonds in $^{32}SO_4^{2-}$ and $^{34}SO_4^{2-}$. In the chemical reduction of sulphate the ratio of the rate constants for the reduction of $^{32}SO_4^{2-}$ and $^{34}SO_4^{2-}$ is about 1.022 which results in a $\delta^{34}S$ of -22‰ for H_2S with respect to the starting sulphate (16).

Biological sulphate reduction also leads to enrichment in ^{32}S in sulphide but the extent of enrichment varies considerably. In so-called 'open' systems, where sulphate is essentially unlimited, $\delta^{34}S$ of sulphide with respect to sulphate approximates -14‰ at high rates of reduction per cell (6,22). As the reduction rate decreases, however, isotopic fractionation may increase for reasons which are not entirely clear: $\delta^{34}S$ values for sulphide of up to -45‰ have been recorded at low rates of biological sulphate reduction (22). Factors such as temperature and electron donor which influence the rate of sulphate reduction may also indirectly affect the degree of isotopic fractionation.

Considerable variability in the isotopic composition of sulphides will also be expected in 'closed' systems where a significant proportion of the total sulphate is metabolized and, as a result, the sulphate becomes progressively heavier due to preferential reduction of ^{32}S. $\delta^{34}S$ values of sulphides precipitated at various stages of reduction will, therefore, reflect the isotopic composition of the intermediate rather than the starting sulphate and will become positive with respect to the latter in the terminal stages of reduction.

The rates of sulphate reduction in sediments are likely to fluctuate widely and the interstitial waters exhibit varying degrees of exchange with the overlying waters. It might be expected, therefore, that the isotopic composition of sulphides precipitated within the sediments will be highly variable and indeed this has been observed in sediments such as those of Southern California where $\delta^{34}S$ values with respect to ocean sulphate of -9 to -62‰ have been reported (21).

The variability of biological isotopic fractionation may also provide an explanation for the wide spread of $\delta^{34}S$ values characteristic of sedimentary deposits although purely chemical explanations are theoretically possible where hydrothermal solutions are involved (30). More confidence can perhaps be placed in data relating the isotopic compositions of co-existing sulphur minerals within deposits. The sulphides of the Kupferschiefer deposit, for example, are highly enriched in ^{32}S, the majority of $\delta^{34}S$ values lying within the range —10 to —40‰ with respect to meteoritic sulphur (26). Comparisons of the $\delta^{34}S$ values of sulphides with those of co-existing sulphates through vertical profiles at various locations of the deposit show a fairly constant differential in any particular profile with $\Delta(\delta^{34}S)$ for sulphate-sulphide ranging from about 10 to 50‰ between locations. These and other observations are consistent with a biological origin for the Kupferschiefer sulphide. The locational differences in the $\Delta(\delta^{34}S)$ sulphate-sulphide values can be explained in terms of the degrees to which the waters of various areas of the depositional environment equilibrated with those of the open ocean. With other deposits, however, the situation is more complex. Isotopic analysis of the sulphide minerals of the Rammelsberg deposits in Germany (1) and of McArthur River (38) indicate separate origins for the sulphides of the lead and zinc minerals as distinct from pyrite: only the latter appears to be of biological origin.

Most strata-bound sulphide deposits are thought to have arisen in marine environments. The isotopic composition of ocean sulphate has fluctuated considerably over geological time (17). Sangster (35) has drawn attention to the fact that the average isotope composition of strata-bound sulphides of different ages appears to fluctuate in a parallel manner to that of ocean sulphate, the sulphides on average being enriched in ^{32}S to the extent of about 14 to 17‰. These values are remarkably close to that found during fast bacterial reduction of sulphate.

A similar parallel also exists between ocean sulphate and organic sulphur in petroleum (36). Since there are good reasons to suspect that the latter was derived from biological sources Sangster suggests that reduction of ocean sulphate was also the principal mechanism by which the sulphide of strata-bound deposits arose, at least since the Cambrian, the earliest period considered by Sangster. Information on Precambrian sulphur minerals is limited. Nevertheless, on the basis of the isotopic composition of some Precambrian evaporitic sulphates, Vinogradov (50) concluded that a complete biological sulphur cycle developed between about 3 and 3.4×10^9 years ago.

FORMATION OF SULPHUR DEPOSITS

A biogenic origin for certain massive deposits of elemental sulphur has also been proposed. A geological feature of native sulphur in sedimentary rocks is the spatial proximity to oil or gas-bearing regions (23). The deposits are also often overlain by 'cap rock' consisting chiefly of gypsum or anhydrite together with calcite. Ivanov (19) has noted the frequent presence of sulphate-reducing bacteria and hydrogen sulphide in oil-bearing mineralized ground waters associated with sulphur deposits. Thus bacterial reduction of gypsum supported by oxidation of hydrocarbons or their organic precursors offers a plausible explanation for the origin of elemental sulphur. This notion is supported by isotopic evidence. As in the case of sedimentary sulphide deposits, sulphur deposits exhibit a spread in $\delta^{34}S$ values (ranging from +40 to —50‰ with respect to meteoritic (20)) similar to that which is characteristic of biogenic sulphides. Moreover, in individual deposits elemental sulphur may be enriched in ^{32}S by up to 50‰ compared to the co-existing sulphate (12), a result which is consistent with a biogenetic relationship between the two sulphur species.

As far as we are aware, however, the sulphate-reducing bacteria are unable to produce elemental sulphur from sulphate and an additional mechanism for the oxidation of hydrogen sulphide is required. One obvious possibility is an inorganic reaction with atmospheric oxygen. In studies on the formation of native sulphur in the Shor-Su and other Russian deposits, however, sulphur-oxidizing bacteria of the thiobacillus type were observed to develop when hydrogen sulphide-rich anaerobic ground waters merged with aerobic surface waters (19,39). The suggestion was made that the oxidizing bacteria may play an important role in sulphur deposition. Sulphur formation in certain Cyrenaican lakes appears to be catalyzed by the photosynthetic bacteria, *Chromatium* and *Chlorobium* (5) which oxidize hydrogen sulphide arising either from bacterial sulphate reduction or from sulphidic hot springs which feed the lakes. About 200 tons per annum of elemental sulphur is produced in Lake Ain-ez-Zauia.

OXIDATIVE WEATHERING

The role of micro-organisms in weathering and reoxidation of reduced sulphides is conjectural. Although bacteria are known to attack many reduced inorganic sulphur minerals (37), the quantitative importance of these reactions in relation to purely chemical oxidation is hard to assess. We do not yet know of any specific features, such as isotopic discrimination, which could differentiate between the end results of biological and chemical oxidation. Speculations on the role of oxidizing bacteria in regeneration of sulphate rely mainly on the virtually ubiquitous association of sulphur-oxidizing bacteria with sulphurous environments and the ability of these organisms to promote active sulphur and sulphide oxidation under laboratory conditions (46).

In the Russian sulphur deposits referred to above intensive sulphur oxidation with the formation of acidic waters has been observed in the aerobic zones (19,24). These zones are also rich in organisms of the thiobacillus type and there seems little doubt that biological oxidation of sulphur plays at least some part in acid formation in these situations. Biological oxidation of metal sulphides is now well documented (27,37). The organisms involved oxidize both sulphur and ferrous ions and their principal function appears to be to generate an acidic ferric liquor which reacts chemically with metal sulphides to generate elemental sulphur and ferrous ions (19).

$2Fe^{3+} + MS \rightarrow M^{2+} + S^0 + 2Fe^{2+}$ (where MS = metal sulphide) Chemical
$2Fe^{2+} + \frac{1}{2}O_2 + 2H^+ \rightarrow 2Fe^{3+} + H_2O$ Biological
$S^0 + \frac{3}{2}O_2 + H_2O \rightarrow H_2SO_4$ Biological

This process has been exploited commercially in the recovery of metals from low-grade sulphide minerals (2) but the extent to which it occurs in the natural environment is unknown.

Acknowledgements. The Baas Becking Geobiological Laboratory is supported by the Bureau of Mineral Resources, the Commonwealth Scientific and Industrial Research Organization and the Australian Mineral Industries Research Association Limited.

REFERENCES

(1) Anger, G., Nielsen, H., Puchelt, H. and Ricke, W. (1966). *Econ. Geol.*, **61**: 511.
(2) Anon. (1958). *Eng. Mining J.*, **159**: 89
(3) Baas Becking, L. G. M. and Moore, D. (1961). *Econ. Geol.*, **56**: 259.
(4) Brown, M. C. Bureau of Mineral Resources Record (in preparation).
(5) Butlin, K. R. and Postgate, J. R. (1954). 'The microbiological formation of sulphur in Cyrenaican lakes' in Cloudsley-Thompson, J. L. (ed.), *Biology of Deserts.* Institute Biology, London, p. 112.
(6) Chambers, L. A. and Trudinger, P. A. (1973) (unpublished observations).
(7) Day, F. H. (1963). *The Chemical Elements in Nature.* George G. Harrap & Co. Ltd, London.
(8) Deuser, W. G. (1971). *Deep-Sea Res.*, **18**: 995.
(9) Erhlich, H. L. and Fox, S. I. (1967). *Biotech. Bioeng.*, **9**: 471.
(10) Eriksson, E. (1960). *Tellus.*, **12**: 63.
(11) Eriksson, E. (1963). *J. Geophys. Res.*, **68**: 4001.
(12) Feely, H. W. and Kulp, J. L. (1957). *Bull. Amer. Assoc. Pet. Geol.*, **41**: 1802.
(13) Freney, J. R. (1967). 'Sulfur-containing organisms' in McLaren, A. D. and Peterson, G. H. (eds), *Soil Biochemistry.* Marcel Dekker Inc., New York.
(14) Freney, J. R., Barrow, N. J. and Spencer, K. (1962). *Pl. Soil*, **17**: 295.
(15) Freney, J. R. and Swaby, R. J. (1974). *Sulphur in Australasian Agriculture.* McLachlan, K. D. (ed.), Sydney University Press, Sydney, p. 31.
(16) Harrison, A. G. and Thode, H. G. (1957). *Trans. Faraday Soc.*, **53**: 1648.
(17) Holser, W. T. and Kaplan, I. R. (1966). *Chem. Geol.*, **1**: 93.
(18) Hutchinson, G. E. (1957). *A Treatise on Limnology.* Vol. I. John Wiley & Sons Inc., New York.
(19) Ivanov, M. V. (1968). *Microbiological Processes in the Formation of Sulfur Deposits.* Israel Programme for Scientific Publications.
(20) Jensen, M. L. (1962). 'Biogenic sulfur and sulfide deposits', in Jensen, M. L. (ed.), *Biogeochemistry of Sulfur Isotopes.* Proc. Nat. Sci. Found. Symp., Yale University.
(21) Kaplan, I. R., Emery, K. O. and Rittenberg, S. C. (1963). *Geochim. Cosmochim. Acta*, **27**: 297.
(22) Kaplan, I. R. and Rittenberg, S. C. (1964). *J. Gen. Microbiol.*, **34**: 195.
(23) Krauskopf, K. B. (1967). *Introduction to Geochemistry.* McGraw-Hill Book Co., New York, p. 276.
(24) Kuznetsov, S. I., Ivanov, M. V. and Lyalikova, N. N. (1963). *Introduction to Geological Microbiology.* McGraw-Hill Book Co., New York.
(25) Lambert, I. B. (1973). *J. Geol. Soc. Aust.*, **20** (in press).
(26) Marowsky, G. (1969). *Contr. Mineral. Petrol.*, **22**: 290.
(27) Miller, J. D. A. (ed.) (1971). *Microbial Aspects of Metallurgy.* Medical and Technical Publishing Co. Ltd, Aylesbury, England.
(28) Miller, L. P. (1950). Contrib. Boyce Thompson Inst. **16**: 85.
(29) Muir, M. (1973). (personal communication).
(30) Ohmoto, H. (1972). *Econ. Geol.*, **67**: 551.
(31) Postgate, J. R. (1959). *Annu. Rev. Microbiol.*, **13**: 505.
(32) Riley, G. A. (1944). *Am. Sci.*, **32**: 132.
(33) Ross, D. A., Degens, E. T. and MacIlvaine, J. (1970). *Science*, **170**: 163.
(34) Roy, A. B. and Trudinger, P. A. (1970). *The Biochemistry of Inorganic Compounds of Sulphur.* Cambridge University Press.

(35) Sangster, D. F. (1968). *Proc. Geol. Assoc. Can.*, **19**: 79.
(36) Sangster, D. F. (1971). *Soc. Mining Geol., Japan.*, Spec. Issue, **3**: 295.
(37) Silverman, M. P. and Ehrlich, H. L. (1964). *Adv. Appl. Microbiol.*, **6**: 153.
(38) Smith, J. W. and Croxford, N. J. W. (1973). *Nature* (London) (in press).
(39) Sokolova, G. A. and Karavaiko, G. I. (1968). *Physiology and Geochemical Activity of Thiobacilli.* Israel Programme for Scientific Translations Ltd.
(40) Sorokin, Yu. I. (1962). *Microbiology* (Eng. trans.), **31**: 329.
(41) Sorokin, Yu. I. (1964). *J. Cons. Perm. Int. Explor. Mer.*, **29**: 41.
(42) Sorokin, Yu. I. (1964). *Int. Rev. Hydrobiol.*, **49**: 307.
(43) Sorokin, Yu. I. (1970). *Arch. Hydrobiol.*, **66**: 391.
(44) Sorokin, Yu. I. (1970). *Oceanology* (Eng. trans.), **10**: 37.
(45) Thode, H. G. (1964). *Bull. Can. Pet. Geol.*, **12**: 246.
(46) Trudinger, P. A. (1967). *Rev. Pure Appl. Chem.*, **17**: 1.
(47) Trudinger, P. A. and Chambers, L. A. (1973). *Geochim. Cosmochim. Acta*, **37**: 1775.
(48) Trudinger, P. A., Lambert, I. B. and Skyring, G. W. (1972). *Econ. Geol.*, **67**: 1114.
(49) Tudge, A. P. and Thode, H. G. (1950). *Can. J. Res.*, **28B**: 567.
(50) Vinogradov, V. I. (1972). *Chem. Geol.*, **10**: 99.

The Chemical Nature of Sulphur Compounds in Soils

C. H. WILLIAMS

Sulphur occurs in soils in both organic and inorganic forms but the proportion of organic to inorganic sulphur may vary widely according to soil type and depth in the profile.

Whilst it has so far proved possible to identify with some degree of certainty the chemical nature of most of the inorganic forms occurring in soils the same cannot be said of the organic sulphur. At the present time only a broad grouping of organic sulphur compounds is possible, based on their reactivity with various reducing agents.

INORGANIC SULPHUR

Inorganic sulphur may occur in soils as sulphate and as compounds of lower oxidation state such as sulphide, polysulphide, sulphite, thiosulphate and elemental sulphur. In well drained, well aerated, soils most of the inorganic sulphur normally occurs as sulphate and the amounts of reduced sulphur compounds are generally less than 1 ppm (22).

Sulphides and compounds of lower oxidation state than sulphate

Considerable amounts of sulphides, and even elemental sulphur can be formed under anaerobic conditions (10,20,29,30,32,60)—particularly in tidal swamps, and poorly drained or waterlogged soils. The greater proportion of these compounds are usually sulphides (10) and in poorly drained soils, they usually occur mainly in the subsoil at depths below the water-table. In such soils the establishment or restoration of aerobic conditions leads to rapid oxidation of sulphide to sulphate and this frequently leads to acidity problems (20,32,55, 60). Sulphides formed under waterlogged conditions can, at times, lead to toxicity problems in rice crops (29,54).

Accurate determination of the amounts of these reduced forms of inorganic sulphur in soils is rarely possible, partially because of the ease with which they can be oxidized on exposure to air, but mainly because of limitations of the analytical procedures available. Smittenberg *et al.*, (52) have proposed methods which have been widely used. These involve digestion of the soil with hydrochloric acid for the determination of monosulphidic sulphur compounds and digestion with tin and hydrochloric acid for the determination of total oxidizable sulphur. The first of these will not release acid-insoluble metal sulphides, and so tends to underestimate the amount of sulphide. The second releases also some sulphur from organic sulphur and thus overestimates the amounts present. Whilst these methods may prove adequate where the amounts of oxidizable sulphur are relatively large they are unsatisfactory for use with soils where the amounts of reduced inorganic sulphur compounds are very small, especially with surface soils containing appreciable amounts of organic sulphur (45).

Sulphate

Sulphate may occur in soils as water-soluble salts, adsorbed by the soil colloids, or as insoluble forms.

Water-soluble sulphate

The amounts of water-soluble sulphate in soils vary widely both between soil types and within the soil profile. In the surface horizons of most well-drained soils they are generally small and appreciable quantities are likely only under conditions of aridity or poor drainage. Considerable seasonal fluctuations in the amounts of soluble sulphate in the surface soil occur, the result of the interaction of seasonal conditions on the mineralization of organic sulphur, leaching of the soluble sulphate, and its uptake by plants (48,62). The sulphate content of soils may also be affected by the application of fertilizers and the sulphate content of rain and irrigation waters.

Appreciable amounts of soluble sulphate are often found in subsoil horizons and the occurrence of free gypsum in the deeper horizons of many semi-arid soils is a well-known pedological feature.

Extraction with water, sodium salts, lithium chloride or calcium chloride solution have been used for the determination of soluble sulphate. Calcium chloride (64) is frequently preferred because of dispersion difficulties with aqueous extraction and because of a lower solubility of organic matter in this reagent. These extracts usually contain some organic sulphur which is reducible by hydriodic acid and steps must be taken to avoid errors from this source if methods involving reduction with this reagent are used in the analytical procedure (24).

Adsorbed sulphate

Soils vary widely in their capacity to adsorb sulphate. Many possess little or no sulphate adsorption capacity but in others sulphate adsorption plays an important part in the retention of sulphate against leaching and in determining its distribution in the profile.

Sulphate adsorption is readily reversible and the amount of sulphate adsorbed is concentration-dependent (12,15,39). The capacity of soils to adsorb sulphate depends upon a number of soil properties the most important of which are clay content (19,46) pH (33,67), the exchangeable cations (16) and the nature of the clay minerals (9,19,39). Adsorption appears to be negligible above pH 6.5 (39,67) and increases with decreasing pH below this (17,39,67). Sulphate adsorption is greater in soils containing large amounts of aluminium and iron oxides (9,13,14,17,19), aluminium oxides generally being more important (14,17,19). Aylmore, Karim and Quirk (1) have shown that the adsorption of sulphate by soil constituents occurs at a number of energetically different reaction sites and that sulphate adsorbed on kaolinite clay is weakly held whereas that adsorbed on iron and aluminium oxides is more tightly bound and may be highly resistant to leaching. The amounts of adsorbed sulphate present in soils may be affected by annual rainfall and by the amounts of sulphate received in rainwater (4,34).

Sulphate adsorption capacity of soils frequently increases with depth in the profile and, partially as a consequence, so too does the amount of adsorbed sulphate. As might be expected sulphate adsorption capacity is affected by pedological factors such as rainfall, drainage and the parent material (6).

Adsorbed sulphate is readily displaced by phosphate and most methods for its determination involve the use of phosphate solutions, most commonly calcium dihydrogen phosphate solutions. As with soluble sulphates these reagents displace some organic sulphur which is reducible by hydriodic acid and appropriate steps must be taken to avoid errors from this source. With soils of high phosphate sorption capacity care must also be taken to ensure that the concentration of phosphate used is adequate to displace all of the adsorbed sulphate.

Insoluble sulphates

Several forms of insoluble sulphates are likely in soils, including barium and strontium sulphates, sulphate associated with calcium carbonate and basic iron and aluminium sulphates

Sulphate occurring in calcareous soils as a co-crystallized impurity in the calcium carbonate (11,67,69) is probably the most common of these and in some soils may account for a high proportion (up to 95 per cent) of the total soil sulphur (67,68). In the subsoil horizons of calcareous soils it frequently accounts for 40 to 50 per cent of the total sulphur.

This fraction of the soil sulphur is determined by extraction with hydrochloric acid (67). The presence of barium in the calcium carbonate whether directly combined with sulphate or not, or the presence of other forms of acid-soluble barium leads to errors resulting in an underestimate of the amounts of sulphate occurring in this form (69).

The small amounts of barium and strontium sulphate present in most soils usually make positive identification impossible. However, Beattie and Haldane (8) have identified barytes occurring as small concretions in certain soils.

Basic iron sulphate, jarosite, $K\,Fe_3(OH)_6(SO_4)_2$ and coquinbite, $Fe_2(SO_4)_3.5\,H_2O$ have been identified in separates taken from tidal marsh soils (18,20, 47) but it seems unlikely that these minerals could be formed in well aerated soils. Basic aluminium sulphates can readily be formed under laboratory conditions when aluminium hydroxide is precipitated in the presence of sulphate ions (7,36,38,49). Basalunite, $Al_4(OH)_{10}SO_4.5H_2O$, is readily formed when the precipitation is carried out in the presence of clay minerals (49,50,51) and has been isolated from deposits occurring in weathered sandstones (35). It would thus seem that basic aluminium sulphate such as this could be formed in natural soils.

ORGANIC SULPHUR

Little is known of the chemical nature of organic sulphur in soil at the present time. The amino acids cystine and methionine have been isolated from soil in trace amounts and their related compounds cysteic acid, methionine sulphoxide and methionine sulphone have been identified in soil hydrolysates. Early analyses understimated the amounts of these forms of soil sulphur due to losses during hydrolysis but recent analyses in which steps were taken to prevent these losses (28) showed that amino acid sulphur accounted for 21 and 30 per cent of the total organic sulphur in two podzolic soils. Approximately 60 per cent of the amino acid sulphur was cystine sulphur.

Although the chemical nature of the organic sulphur cannot be defined with the same precision as is possible for inorganic sulphur, several broad groups have been recognized.

1. Organic sulphur which is reduced to H_2S by hydriodic acid. This sulphur is not bonded directly to carbon and is believed to be largely in the form of ester sulphates (22).
2. Organic sulphur which is not reduced by hydriodic acid. This is believed to comprise all of the sulphur which is bonded directly to carbon.
3. Organic sulphur which is reduced to inorganic sulphide by Raney nickel and which appears to consist almost entirely of sulphur in the form of amino acids (26).

No satisfactory direct method is available for the determination of total organic sulphur or of the organic sulphur reducible by hydriodic acid. These must either be determined on the soil residue after extraction of inorganic sulphur or by difference involving the subtraction of total inorganic sulphur from the total sulphur or total hydriodic acid reducible sulphur. Both are thus subject to error in soils which contain inorganic sulphates insoluble in phosphate solution or hydrochloric acid and in soils containing barium ions soluble in either of these reagents.

The possibility of such errors is indicated in Figure 2-3 in which the total organic sulphur for a krasnozem is plotted against total nitrogen. While the organic sulphur in the surface 45 cm appears to be closely related to nitrogen, soil depths below 45 cm contained considerably more sulphur relative to nitrogen. This suggests that in the subsoil, either the organic matter has a much lower N:S ratio or that the presence of insoluble inorganic sulphate has led to an over-estimate of the organic sulphur. The total nitrogen may include small amounts of fixed ammonium so that the ratio organic N:S in the subsoil horizons would be even lower. Extraction of the 60 to 75 cm sample of this soil, after phosphate extraction, with neutral ammonium fluoride, 0.1 N sodium hydroxide, and 6 N hydrochloric acid dissolved additional inorganic sulphate in each case but in no case was this sufficient to account for more than 30 per cent of the apparent discrepancy.

Total organic sulphur

Sulphur is closely associated with carbon and nitrogen in the soil organic matter and similar ratios between these three elements have been reported (C:N:S about 140:10:1.3), mostly for surface soils, from widely separated parts of the world (2,53,59). There is a tendency for this ratio to be maintained when organic matter is lost from the soil as a result of cultivation (31,64) and when organic matter is accumulated in the soil under

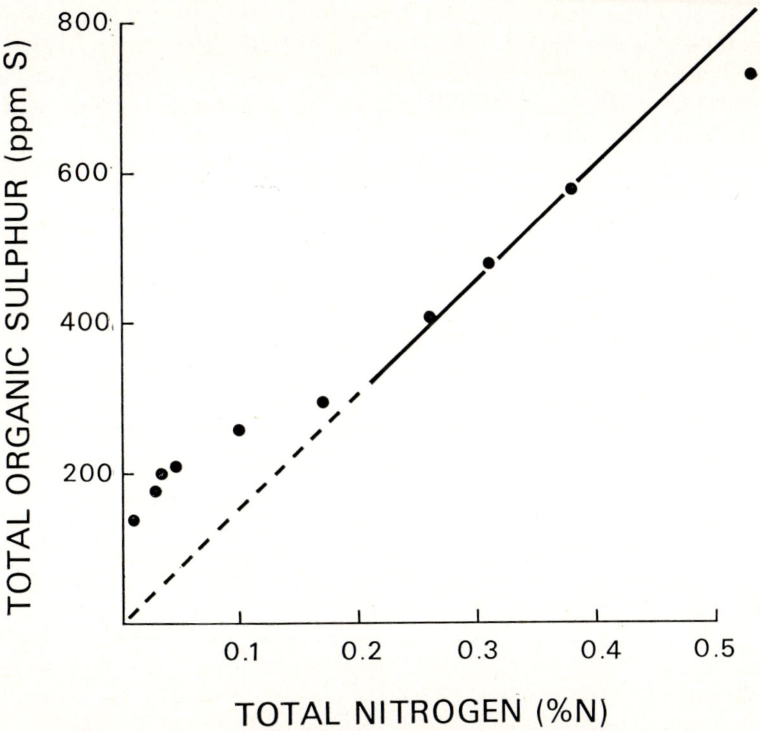

Figure 2-3 The relationship between total nitrogen and the difference between total sulphur and sulphate extracted by calcium dihydrogen phosphate solution (total organic sulphur) in a krasnozem from Berry, N.S.W.

leguminous pasture (5,37,63). However, substantial changes, especially in the C:S ratio have been reported by Walker, Thapa and Adams (58) following the establishing of clover-grass pasture on soils previously carrying manuka scrub.

Although the mean C:N:S ratios for various groups of soils are similar, appreciable variations in ratio occur between individual soils, while smaller variations occur between soil groups. Differences between soil groups have been attributed to parent material (56,65,69) and other soil-forming factors (57).

Except for soils with organic matter accumulation in the B-horizon and soils formed under conditions where erosion-deposition cycles have led to buried profiles, organic sulphur tends to decrease with depth in the profile.

Hydriodic acid-reducible sulphur

In general the hydriodic acid-reducible sulphur (ester sulphate) fraction accounts for approximately 50 per cent of the total organic sulphur, although analyses covering wide ranges of soils have indicated that the percentage may range from 30 to 70 per cent (22,25,41,44,53,66). In some soils the percentage of this fraction in the total organic sulphur tends to remain fairly constant throughout the soil profile but in others the percentage increases with increasing depth in the profile (Williams, unpublished data). In Iowa soils, Tabatabai and Bremner (53) have found that the percentage of this fraction in the total soil sulphur increased with increase in sample depth. Whether this increase in the proportion of hydriodic acid-reducible sulphur with depth reflects an actual change in the chemical nature of the organic sulphur or is due to errors in the chemical method for its determination, as discussed above, is not certain.

Extraction of soils with sodium bicarbonate at pH 10 and with chelating resin (24) has indicated that much of the hydriodic acid-reducible sulphur occurs in the high molecular weight (humic acid) fraction of the soil organic matter.

Carbon-bonded sulphur

This fraction represents the difference between the total organic sulphur and the hydriodic acid-reducible fraction and so comprises, in general, approximately half of the total organic sulphur. In some soil profiles at least the carbon-bonded sulphur is closely related to nitrogen suggesting that there is little change in the chemical nature of this fraction throughout the profile. However, the ratio N:carbon-bonded S has been found to range from < 10 to 20 (Williams, unpublished data). In general the relationship of this fraction with nitrogen appears to be much closer than is the case with hydriodic acid-reducible fraction.

This fraction accounted for the greater part (84 per cent) of the soil organic sulphur accumulated under clover pasture (Table 2-3) and of the organic sulphur lost (81 per cent) as a result of cereal cropping. Of the carbon-bonded sulphur accumulated under pasture 42 per cent of it was reducible by Raney-nickel.

Raney-nickel-reducible sulphur

This fraction is a sub-fraction of the carbon-bonded sulphur and probably represents the cystine and methionine sulphur in the soil organic matter (26). In a range of Australian soils (25) it has been found to account for an average of 56 per cent (45 to 93) of the total carbon-bonded sulphur and 30 per cent (22 to 54) of the total soil sulphur. Lowe and DeLong (44) and Lowe (41,42) have reported values for Canadian soils indicating that this fraction may account for 46 to 58 per cent of the total sulphur in organic soils and 12 to 32 per cent in mineral soils while Tabatabai and Bremner (53) reported a mean value of 11 per cent of the total sulphur in thirty-seven Iowa soils. The original method of Lowe and DeLong (43) for the determination of Raney-nickel-reducible sulphur however, is subject to serious interference from iron and manganese (25) leading to low values. The Canadian and American values thus could be low in some cases and should be regarded as minimum values only.

Organic sulphur reduced by tin and hydrochloric acid

The procedure of Smittenberg *et al.* (52) involving the digestion of soils with tin and hydrochloric acid was intended primarily for the determination of reduced forms of inorganic sulphur in soils, although the authors indicated that it also reduced some organically bound sulphur. In well-aerated soils,

TABLE 2-3

Sulphur fractions, ppm S, in matched pairs of soils taken from unfertilized sites and adjacent improved pasture or cropped sites. The pasture site had been under subterranean clover for 40 to 50 years and the cropped sites (mean of ten) had been cropped to wheat for 50 to 60 years

Sulphur fraction	Pasture soils*			Cropped soils**		
	Native pasture	Sub-clover pasture	Accumulation under pasture	Uncultivated	Cultivated	Loss due to cropping
Total S	109	252	143	131	94	37
HI-reducible S	49	72	23	56	49	7
Carbon-bonded S	60	180	120	75	45	30
Raney Ni-reducible S	40	91	51			

* Data from Freney, Melville and Williams (26).

** Data from K. D. McLachlan, personal communication.

especially surface soils, the amounts of reduced forms of inorganic sulphur compounds are considered to be very low, so that, in such soils, most of the sulphur reduced by this treatment is probably organic.

In twelve Australian surface soils, Melville, Freney and Williams (45) found that sulphur accounting for an average of 2.8 per cent of the total sulphur was reduced by this procedure. The fraction was highly correlated with total soil sulphur, total organic sulphur and total nitrogen and, in a single profile, the tin and hydrochloric reducible sulphur decreased with depth and was again highly correlated with total organic sulphur and total nitrogen.

The reaction of model compounds with tin and hydrochloric acid (45) indicated that the C=S bond is particularly susceptible to reduction by this procedure, but the reaction did not quantitatively reduce this bond in all compounds. The reaction also partially reduced the sulphur in S-sulphocysteine. Thus the nature of the organic sulphur in soils reduced by this procedure is uncertain.

AVAILABILITY OF SOIL SULPHUR FRACTIONS TO PLANTS

Apart from fully available water-soluble sulphate, adsorbed sulphate has been shown to be highly available and to comprise a major source of plant-available sulphur in many soils (3,21,27,34,68). The ready reversibility of sulphate adsorption would seem likely to ensure this. In addition to the adsorbed sulphate in the surface horizons, subsoil sulphate also frequently plays an important part in providing available sulphur for deep-rooted plant species (40).

Barrow (4) has shown in pot culture that adsorbed sulphate has lower availability to plants in soils of high adsorption capacity than in soils of low adsorption capacity. However, under field conditions this would seem likely to prove of little consequence and any slight disadvantage arising from this would be fully offset by the advantages stemming from the retention of available sulphate against leaching.

Of the insoluble sulphates, that associated with calcium carbonate has been shown to be quite unavailable to plants (68). However, the solubility of barium sulphate (solubility product approximately 1×10^{-10}) is high enough to enable finely divided material to contribute some available sulphur should the concentration of sulphate in the soil solution become low enough (say < 0.3 ppm S), but it could be expected that nodular forms of barium sulphate (8) would be totally unavailable. Basic iron and aluminium sulphates could also be expected to have extremely low plant-availability.

It is generally accepted that plants take up their sulphur in the sulphate form and thus the availability of soil organic sulphur depends upon its mineralization. This will depend upon climatic factors such as temperature and moisture (61), which affect the rate of mineralization, as well as the chemical nature of the soil organic matter itself. Whilst laboratory and glasshouse experiments have indicated that plant-available sulphur is derived from both the hydriodic acid-reducible and the carbon-bonded fractions of the organic sulphur (23,26) it is clear that a more detailed chemical subdivision of both fractions will be necessary before organic fractions directly involved in the mineralization of the organic sulphur can be adequately identified.

REFERENCES

(1) Alymore, L. A. G., Karim, M. and Quirk, J. P. (1967.) *Soil Sci.*, **103**: 10.
(2) Barrow, N. J. (1961). *Soils and Fertilizers*, **24**: 169.
(3) Barrow, N. J. (1967). *Soil Sci.*, **104**: 242.
(4) Barrow, N. J. (1969). *Soil Sci.*, **108**: 193.
(5) Barrow, N. J. (1969). *Aust. J. Exp. Agr. Anim. Husb.*, **9**: 437.
(6) Barrow, N. J., Spencer, K., and McArthur, W. M. (1969). *Soil Sci.*, **108**: 120.
(7) Bassett, H. and Goodwin, T. H. (1949). *J. Chem. Soc.*, **1949**: 2239.
(8) Beattie, J. A. and Haldane, A. D. (1958). *Aust. J. Sci.*, **20**: 274.
(9) Berg, W. A. and Thomas, G. W. (1959). *Proc. Soil Sci. Soc. Am.*, **23**: 348.
(10) Brummer, G., Grunwaldt, H. S. and Schroeder, D. (1971). *Z. PflErnähr Dung. Bodenk.*, **128**: 208.
(11) Brummer, G., Grunwaldt, H. S. and Schroeder, D. (1971). *Z. PflErnähr Dung. Bodenk.*, **129**: 92.
(12) Chang, M. L. and Thomas, G. W. (1963). *Proc. Soil Sci. Soc. Am.*, **27**: 218.
(13) Chao, T. T., Harward, M. E. and Fang, S. C. (1962). *Proc. Soil Sci. Soc. Am.*, **26**: 2.
(14) Chao, T. T., Harward, M. E. and Fang, S. C. (1962). *Soil Sci.*, **94**: 276.
(15) Chao, T. T., Harward, M. E. and Fang, S. C. (1962). *Proc. Soil Sci. Soc. Am.*, **26**: 234.
(16) Chao, T. T., Harward, M. E. and Fang, S. C. (1963). *Proc. Soil Sci. Soc. Am.*, **27**: 35.
(17) Chao, T. T., Harward, M. E. and Fang, S. C. (1964). *Proc. Soil Sci. Soc. Am.*, **28**: 632.
(18) Clark, J. S., Gobin, C. A. and Sprout, P. N. (1961). *Can. J. Soil Sci.*, **41**: 218.
(19) Ensminger, L. E. (1954). *Proc. Soil Sci. Soc. Am.*, **18**: 259.
(20) Fleming, J. F. and Alexander, L. T. (1961). *Proc. Soil Sci. Soc. Am.*, **25**: 94.
(21) Fox, R. L., Olson, R. A. and Rhoades, H. F. (1964). *Proc. Soil Sci. Soc. Am.*, **28**: 243.
(22) Freney, J. R. (1961). *Aust. J. Agric. Res.*, **12**: 424.
(23) Freney, J. R. and Swaby, R. J. (1974). *Sulphur in Australasian Agriculture.* McLachlan, K. D. (ed.), Sydney University Press, Sydney, p. 31.
(24) Freney, J. R., Melville, G. E. and Williams, C. H. (1969). *J. Sci. Fd Agric.*, **20**: 440.
(25) Freney, J. R., Melville, G. E. and Williams, C. H. (1970). *Soil Sci.*, **109**: 310.
(26) Freney, J. R., Melville, G. E. and Williams, C. H. (unpublished).
(27) Freney, J. R. and Spencer, K. (1960). *Aust. J. Agric. Res.*, **11**: 339.
(28) Freney, J. R., Stevenson, F. L. and Beavers, A. H. (1972). *Soil Sci.*, **114**: 468.
(29) Green, V. E. (1957). *Agron. J.*, **49**: 468.
(30) Harmsen, G. W. (1954). *Pl. Soil*, **5**: 324.
(31) Harper, H. J. (1959). Okla. Agric. Exp. Sta. Bull. B-536.
(32) Hart, M. G. R. (1959). *Pl. Soil*, **11**: 215.
(33) Harward, M. E. and Reisenauer, H. M. (1966). *Soil Sci.*, **101**: 326.
(34) Hasan, S. M., Fox, R. L. and Boyd, C. C. (1970). *Proc. Soil Sci. Soc. Am.*, **34**: 897.
(35) Hollingworth, S. E. and Bannister, F. A. (1952). *Mineral Mag.*, **29**: 1.
(36) Hsu, P. H. and Bates, R. F. (1964). *Mineral Mag.*, **33**: 749.
(37) Jackman, R. H. (1964). *N.Z. J. Agric. Res.*, **7**: 445.
(38) Johansson, G., Lundgren, G., Sillen, L. G. and Soderquist, R. (1960). *Acta Chim. Scand.*, **14**: 769.
(39) Kamprath, E. J., Nelson, W. L. and Fitts, J. W. (1956). *Proc. Soil Sci. Soc. Am.*, **20**: 463.
(40) Lichtenwalner, D. C., Flenner, A. L. and Gordon, N. E. (1923). *Soil Sci.*, **15**: 157.
(41) Lowe, L. E. (1965). *Canad. J. Soil Sci.*, **45**: 297.
(42) Lowe, L. E. (1969). *Canad. J. Soil Sci.*, **49**: 375.
(43) Lowe, L. E. and DeLong, W. A. (1962). *Canad. J. Soil Sci.*, **42**: 223.
(44) Lowe, L. E. and DeLong, W. A. (1963). *Canad. J. Soil Sci.*, **43**: 151.

(45) Melville, G. E., Freney, J. R. and Williams, C. H. (1971). *Soil Sci.*, **112**: 245.
(46) Neller, J. R. (1959). *Proc. Soil Sci. Soc. Am.*, **23**: 246.
(47) Schwertmann, U. (1961). *Naturwissenschaften*, **48**: 159.
(48) Simon-Sylvestre, G. (1965). *C.r. Hebd. Séanc Acad. Agric. Fr.*, 426.
(49) Singh, S. S. (1967). *Soil Sci.*, **104**: 433.
(50) Singh, S. S. and Brydon, J. E. (1967). *Soil Sci.*, **103**: 162.
(51) Singh, S. S. and Brydon, J. E. (1969). *Soil Sci.*, **107**: 12.
(52) Smittenberg, J., Harmsen, G. W., Quispel, A. and Otzen, D. (1951). *Pl. Soil*, **3**: 353.
(53) Tabatabai, M. A. and Bremner, J. M. (1972). *Soil Sci.*, **114**: 380.
(54) Vamos, R. (1964). *J. Soil Sci.*, **15**: 103.
(55) Walker, P. M. (1972). *Aust. J. Soil Res.*, **10**: 127.
(56) Walker, T. W. and Adams, A. F. R. (1958). *Soil Sci.*, **85**: 307.
(57) Walker, T. W. and Adams, A. F. R. (1959). *Soil Sci.*, **87**: 1.
(58) Walker, T. W., Thapa, B. K. and Adams, A. F. R. (1959). *Soil Sci.*, **87**: 135.
(59) Whitehead, D. C. (1964). *Soils and Fertilizers*, **27**: 1.
(60) Wiklander, L. and Hallgreen, G. (1949). *K. LantbrHögsh. Annlr*, **16**: 811.
(61) Williams, C. H. (1967). *Pl. Soil*, **26**: 205.
(62) Williams, C. H. (1968). *Aust. J. Soil Res.*, **6**: 131.
(63) Williams, C. H. and Donald, C. M. (1957). *Aust. J. Agric. Res.*, **8**, 179.
(64) Williams, C. H. and Lipsett, J. (1961). *Aust. J. Agric. Res.*, **12**: 612.
(65) Williams, C. H. and Steinbergs, A. (1958). *Aust. J. Agric. Res.*, **9**: 483.
(66) Williams, C. H. and Steinbergs, A. (1959). *Aust. J. Agric. Res.*, **10**: 340.
(67) Williams, C. H. and Steinbergs, A. (1962). *Pl. Soil*, **17**: 279.
(68) Williams, C. H. and Steinbergs, A. (1964). *Pl. Soil*, **21**: 50.
(69) Williams, C. H., Williams, E. G. and Scott, N. M. (1960). *J. Soil Sci.*, **11**: 334.

Sulphur Transformations in Soils

J. R. FRENEY and R. J. SWABY

The transformations of sulphur in soil are many and varied and often the changes are cyclic; the element changes from inorganic to organic forms and back again due to the presence of living organisms.

During soil formation the sulphides of primary minerals are converted to sulphate which is changed in turn to many and varied organic compounds by micro-organisms, plants and animals, and thus sulphur can be found in soils in many forms. In the surface horizon of well-drained non-calcareous soils of the humid regions most of the sulphur accumulates in organic forms, whereas in arid zone and calcareous soils large amounts of inorganic sulphate accumulate, and under waterlogged conditions appreciable sulphide is found.

This review will cover mainly the period since 1966 as work prior to that date has already been reviewed (8,17,22,27,46,56,58).

IMMOBILIZATION

The introduction of productive legumes in southern Australian pastures and the continual application of superphosphate together with other deficient nutrients has resulted in the accumulation of organic matter in these soils (7,61). For example, in a Coolup sand, sulphur was incorporated into organic forms along with nitrogen and organic phosphorus to give C:N:S:P ratios of 188:10:1.2:0.47 (7). These elements accumulated at the rate of 440, 38, 4.4 and 1.7 kg/ha/year in the top 12.5 cm of soil. Only one-quarter of the sulphur applied in fertilizer (20 kg S/ha/year) was converted to organic forms, and this contrasts with the results of Williams and Donald (61) who recovered most of the applied sulphur in the accumulated organic matter. This probably reflects the difference between leaching and non-leaching environments.

Sulphur also accumulated along with organic matter in soil left uncultivated and unfertilized (25,51). Jenkinson (25) found gains for C, N, S and P in organic matter of 560, 49, 6.9 and 5.9 kg/ha/year respectively in the top 22.9 cm of Broadbalk soil where trees and shrubs were removed regularly, and gains of 530, 45, 6.9 and 5.6 kg/ha/year on Broadbalk wooded soil. This gain in sulphur in industrialized England can be attributed to input in rain alone, but sulphur dioxide could also have been absorbed directly from the atmosphere (25).

Larson *et al.* (29) examined the effects on sulphur incorporation of adding different types and amounts of plant residues (alfalfa, cornstalks, sawdust, oat straw, and bromegrass) to soil in a field experiment. Organic C, N, S and P contents of the soils increased in proportion to the amount of plant residues added. Average increases over the control plots after a period of 11 years were 47, 37, 45 and 14 per cent respectively for C, N, S and P for the 16 tons/ha/year treatment. Sawdust resulted in a significantly lower value for N and S than the other residues. The low sulphur content of sawdust undoubtedly explains the slow build up of organic sulphur with the application of this residue.

Factors affecting immobilization

Stotzky and Norman (50) concluded, from studies using glucose as the carbon source, that a C:S ratio of 900 or lower would provide adequate sulphur for maximum microbial activity, and that as most plant residues have a C:S ratio less than 900, sulphur should not be limiting for the decomposition of residues under natural conditions. However, other workers (3,32,49) have shown that immobilization of sulphur will occur during the decomposition of plant residues with C:S ratios well below 900:1. This suggested that the organic sulphur was not readily available for the needs of the micro-organisms, or that the organisms involved have a higher sulphur requirement for the decomposition of plant residues than for glucose. The latter alternative is more likely, because Stewart *et al.* (48), working with pure compounds, showed that micro-organisms required more sulphur for the maximum decomposition of cellulose (1 part S:300 parts C) than for glucose (1 part S:600 to 900 parts C). This higher sulphur requirement may be related to the production of extracellular enzymes which are necessary for the breakdown of cellulose (48).

It is therefore important to consider the adequacy of soil sulphur supplies during the management of crop residues. The decomposition rate of straw in a sulphur deficient soil depended upon the sulphur content of the straw (47). It was necessary for the sulphur content of the straw to exceed 0.15 per cent to obtain the maximum rate of decomposition. Straws with < 0.15 per cent S added to soil depressed wheat growth by immobilizing sulphur, but straws with a higher S content increased yields (47).

Chemical nature of immobilized sulphur

In agricultural practice, where fertilizer sulphate has been applied, some of the sulphur is converted to organic forms by plants and some by micro-organisms alone. Freney *et al.* (19) showed that when 10 ppm SO_4-S was added to soil there was steady incorporation of sulphur into organic matter over a period of 24 weeks up to a maximum of 50 per cent of the added sulphur. Addition of glucose increased the amount of sulphur incorporated up to 82 per cent.

Sterilizing treatments markedly decreased the incorporation of labelled sulphur indicating the important role of soil micro-organisms in this process. Autoclaving the soil before addition of sulphate reduced the incorporation in two weeks from 45.5 to 1.3 per cent; formaldehyde and ethylene oxide reduced incorporation to a lesser extent. These results also suggested that isotopic exchange was not involved to any extent (19). They also showed that labelled sulphur was incorporated into both the 'hydriodic acid-reducible' and the 'carbon-bonded sulphur' fractions, but that the sulphur in the hydriodic acid-reducible fraction was more highly labelled than the sulphur bonded to carbon (19).

Koter *et al.* (28) also studied the incorporation of $^{35}SO_4$-S into soil organic matter at 30°C for three weeks. Peat soils incorporated more labelled sulphur than mineral soils and a larger proportion of the labelled sulphur was in the acid soluble fulvic acid fraction of a sodium hydroxide extract. Other workers (19,38) have also found little label in the humic acid fraction when sodium hydroxide was used as an extractant, but if sodium bicarbonate is used to extract labelled organic matter (19) then most of the labelled sulphur is found in the humic acid fraction. As some organic matter is known to be degraded in sodium hydroxide solutions but not in sodium bicarbonate (18) it would appear that most of the sulphur is incorporated into the high molecular humic acid fraction.

In laboratory experiments in which $Na_2\,^{35}SO_4$ was incubated with cow manure (36), only 16 per cent of the ^{35}S was incorporated into organic compounds in the first month, mainly into the humic acid fraction. At the end of 2 months approximately 44 per cent of the ^{35}S occurred in organic substances with 20 per cent in the humic acid fraction and with significant amounts in the volatile and insoluble fractions.

MINERALIZATION

Factors affecting mineralization

It is believed that the conversion of organic sulphur to inorganic sulphate is mainly carried out by soil micro-organisms (1) and thus any variable which affects the growth of micro-organisms will affect the mineralization of sulphur. Thus temperature, moisture, pH (9,10,59), and availability of food supply (2) have all been shown to affect the mineralization of sulphur.

Mineralization of sulphur was markedly suppressed at 10°C and no mineralization occurred in some soils at this temperature over a 64-day incubation period (59). Mineralization of sulphur increased with increasing temperature from 20°C to 40°C but was less at 50°C than at 40°C (10,59).

At low ($<$ 15 per cent) and high ($>$ 40 per cent) moisture levels mineralization of sulphur was considerably retarded (59). Optimum moisture content for mineralization was 60 per cent maximum water holding capacity (9).

Addition of calcium carbonate increased the amounts of sulphur mineralized. With three soils the amount of sulphur mineralized was directly proportional to pH up to a value of 7.5. Above pH 7.5 mineralization increased more rapidly suggesting that a further factor—possibly chemical hydrolysis—was affecting the process. With three other soils mineralization was proportional to the amount of calcium carbonate added and not to the pH attained (59).

Further evidence that the transformation of organic sulphur to inorganic sulphate is mainly microbial, is given by the effect of toluene and formaldehyde on mineralization. In all cases toluene suppressed mineralization but in some soils formaldehyde only partially suppressed the transformation (59).

In the presence of calcium carbonate, mineralization was decreased by toluene and formaldehyde but in no case was it completely eliminated. This suggests that while the main effect of calcium carbonate addition is to stimulate microbial mineralization, it also promotes chemical reactions such as hydrolysis of organic sulphur (59).

When soils are dried and then remoistened prior to incubation they release appreciably more sulphate than they do when incubated without prior drying (59). Thus care should be exercised in interpreting the results of work on mineralization of sulphur when the soils are dried before incubation, although these results may be more realistic to the field situation after dry summers.

The amount of sulphur mineralized from unamended soils does not appear to be directly related to soil type, to total amounts of carbon, nitrogen or sulphur, to C:S, N:S, or C:N ratios, to soil pH, or to mineralizable N (23, 26,44,54,59).

Sulphur mineralization follows a number of patterns: (a) immobilization of sulphur during the initial stages of incubation, followed by mineralization of sulphur in the later stages (4,6,7,23,33,54); (b) a steady, linear, release with time over the whole period of incubation (59); (c) a rapid release of sulphate during the first few days followed by a slower linear release (59); (d) a rate of release which decreased with time (59). The pattern of release does not appear to be related to any soil property, but is probably due to the chemical nature of the decomposing fraction of the soil organic matter. Better predictions may be obtained when the main sulphur compounds in soil are identified and their turnover rate determined.

Freney and Spencer (21) showed that mineralization of sulphur in the presence of plants was greater than that which occurred in the unsown soil and suggested that the increased mineralization may have been due to the greater proliferation of organisms under plants. This result was subsequently confirmed (12,35), though other workers using the same balance sheet approach have found some soils which behave in this manner and other soils which mineralize less sulphur under plants than in the absence of plants (6,26).

The reasons for these differences are not known and it is difficult to relate them to soil properties because some workers do not present all of the data necessary to make comparisons. There can be no doubt that the estimation of mineralization under plants is subject to large errors unless extreme care is exercised e.g., in the measurement of sulphur inputs in daily watering, the exclusion of atmospheric contamination by sulphur dioxide (12), the complete recovery of roots from the soil (6) and the use of satisfactory analytical methods for the determination of total sulphur and sulphate. For example recoveries of total sulphur from plant material by the nitric and perchloric acid digestion may be as low as 75 per cent of the total sulphur (37).

Seasonal fluctuations in mineral sulphur

A few workers have looked at mineralization of sulphur in field situations. Williams and Lipsett (30,62) studied the changes which had occurred in soils cultivated for wheat in southern New South Wales. Many of the soils had carried at least twenty crops since being brought under cultivation and had lost an average of 30 per cent of their organic matter during this period. The N:S ratio of the organic matter in these soils was 7.3:1 and it had not been changed by the extensive cropping, indicating that in the long-term nitrogen and sulphur were mineralized in proportions similar to those in the original soil organic matter. Workers in Colorado confirmed this result and pointed out that as the critical N:S ratio of young wheat plants is about 17:1, crops grown on these soils without the addition of nitrogenous fertilizer could be expected to derive adequate sulphur from the mineralization of organic sulphur (49).

The mineralization of carbon, nitrogen, and sulphur from bare plots which had been fertilized with different manures was studied over a period of thirty years (39). The three elements decreased exponentially and the annual decrease in organic matter averaged 1.8 per cent. After thirty years, only 57.4 per cent of the initial carbon, 54.7 per cent of the initial nitrogen and 51.6 per cent of the initial sulphur remained in the soil, again showing that nitrogen and sulphur are mineralized at approximately the same rate.

When soils are incubated in the laboratory under aerobic conditions for relatively short periods the ratio for N:S mineralized is generally considerably greater than the N:S ratio of the original soil (4,7,24,40,54,57,59), although in a few cases reported a greater proportion of sulphur relative to nitrogen is mineralized (33,42,44). Some of the reasons for the differences between the field and laboratory situations have already been discussed (22).

Seasonal changes in sulphate sulphur in either bare soil, cultivated fields, or soil under pasture have been studied in a number of locations (5,7,40,41, 42,43,45,60). In soil under subterranean clover pasture (60) sulphate accumulated in the surface soil during summer immediately after senescence of the pasture. High concentrations were maintained throughout the summer-autumn period and these decreased to low values in winter and spring. The higher values in summer probably resulted from mineralization of soil organic sulphur under favourable moisture and temperature conditions and lack of plant uptake. The low values in winter and spring were probably due to leaching, plant uptake and low rates of mineralization at low soil temperatures (60).

Similar seasonal patterns were observed elsewhere (7,40) although Barrow (5) ascribed the higher levels of sulphate in summer to the release of sulphur from soil organic matter by desiccation (4,15,54,59), whereas Williams (60) suggested that only small amounts of sulphate were released in his study by this phenomenon.

In general, fluctuations in mineral sulphur paralleled those of nitrate although Simon-Sylvestre (40) found that at the peak of mineralization, 2.7 per cent of the total nitrogen was mineralized compared with 12.3 per cent for sulphur. The relative amounts of sulphur released are far greater than those reported by other workers and no logical explanation can be advanced for this difference.

Source of mineralizable sulphur

Few attempts have been made to study the dynamics of sulphur in soil, i.e., to evaluate the amount of available sulphur in soil (i.e., soluble plus adsorbed sulphate), the amount of the biologically and chemically active reserve forms, and the rate of turnover which regulates the transition from one form to the other. Fortini *et al.* (14) attempted to do this with ^{35}S labelled sulphur and calculated that the daily rate of regeneration of available sulphate, in a volcanic soil was 2.7 μg S and in an organic soil was 4.6 μg S. In the volcanic soil the physico-chemical and biological transformations occurring daily in total sulphur affected 7.7 per cent of the soluble sulphate, 7.3 per cent of the adsorbed sulphate and 9.4 per cent of the reserve sulphur.

Freney *et al.* (20) also using ^{35}S labelled soil organic matter have attempted to define the source of the available sulphur in soil. Over a plant growth period of nine months most of the available sulphur removed by plants came from the ester sulphate fraction, although there were changes in all sulphur fractions. Most of the change in the carbon-bonded sulphur fraction was due to decomposition of the sulphur-containing amino acids.

Aryl sulphatase activity

Ester sulphate accounts for approximately 50 per cent of the total sulphur in surface soils from Australia, Canada and U.S.A. (16,17,31,55), and it has been suggested that sulphatases may play an important role in the mineralization of organic sulphur in soil.

Soils have been shown to contain enzymes that hydrolyze aryl sulphates (11,52,53) and a method has been developed for the determination of aryl sulphatase activity in soil (52). These workers (53) showed that air drying moist soils caused a marked increase in aryl sulphatase activity, although cycles of drying and wetting reduced aryl sulphatase activity (11). In general, aryl sulphatase activity was found to be weakly correlated with organic matter content (11,53) and activity decreased with depth in the profile as organic matter decreased (53).

Sulphur mineralized from a range of soils from Iowa was not related to aryl sulphatase activity (54). Aryl sulphatase activity in that study, expressed as μg of p-nitrophenol released, from potassium p-nitrophenyl sulphate, per g of soil per hour ranged from 7.1 to 34.2, whereas the sulphur mineralized ranged from 0.7 to 5.5 μg S per g soil per ten weeks (54).

Table 2-4 presents calculations based on the results of Tabatabai and Bremner (54) for two of their soils, and indicates the theoretical amount of sulphur released by aryl sulphatase activity over a period of ten weeks.

The amounts of sulphate which should be released by aryl sulphatase activity in ten weeks are obviously absurd, as the calculated release is greater than the total sulphur content of the soil. On the other hand, if the sulphur mineralized during aerobic incubation is calculated in terms of sulphate sulphur produced per hour, so that a ready comparison can be made between the rates, then it is apparent that aryl sulphatase activity is 1,000-fold greater than mineralization. Even if allowance is made for the immobilization of sulphur during the first two weeks (54) the rate of mineralization of sulphur can only be a small fraction of the aryl sulphatase activity. If sulphate esters are present in soil (16,31,55), then, with this level of activity, sulphur deficient soils should never be found. What then is the reason for this discrepancy?

Aryl sulphatase activity determinations were made under optimum conditions of pH, temperature and substrate concentration, and were carried out over a relatively short time period. The conditions pertaining during the mineralization study at 30° were probably suboptimal for maximum aryl

TABLE 2-4
Comparison of the amounts of sulphur mineralized during aerobic incubation of soils at 30°C with the calculated release due to aryl sulphatase activity* (ppm S)

Soil	Total sulphur	Sulphur mineralized** during		Calculated sulphate sulphur release due to aryl sulphatase activity for	
		1 h	10 weeks	1 h	10 weeks
Hagener	78	.0014	2.3	1.6	2,746
Ida	300	.0010	1.7	7.9	13,227

* The original data and that used for the calculation of sulphate released by aryl sulphatase activity is taken from Tabatabai and Bremner (54).
** Sulphur mineralized during one hour is calculated from the ten-week value and does not take into account the immobilization which occurred at the beginning of the incubation.

sulphatase activity (52). Also conditions which normally pertain in soils in the field environment would be greatly removed from the optimum conditions of the assay.

Even though aryl sulphatase activity was linear up to eight hours in one soil (52) it is most unlikely that this rate would be maintained during the ten weeks of the incubation. Again aryl sulphatase activity was determined in the presence of toluene which would have suppressed microbial interference and probably eliminated immobilization.

Another possibility is that specific sulphatase enzymes do not exist in soil and that the splitting of aryl sulphates was due to non-specific enzymes which occur in soil and which normally are involved in other reactions.

Other possibilities are that aryl sulphates do not occur naturally in soils, the aryl sulphate group is in a position where it cannot be attacked, or that the reaction is inhibited in the natural environment by adsorption on sesquioxides, clays etc. Further work is needed to establish the real reasons for the apparent high aryl sulphatase activity under laboratory conditions and the low sulphate release figures under field conditions or during incubation.

Of the other transformations which occur in soil, sulphate reduction has been reviewed recently by Dommergues and Jacq (13), Kelly (27), and Nicholas (34), and the oxidation of sulphur and sulphide is reviewed elsewhere in this volume.

REFERENCES

(1) Alexander, M. (1961). *Introduction to Soil Microbiology.* John Wiley and Sons Inc., New York.
(2) Barrow, N. J. (1960). *Aust. J. Agric. Res.*, **11**: 317.
(3) Barrow, N. J. (1960). *Aust. J. Agric. Res.*, **11**: 960.
(4) Barrow, N. J. (1961). *Aust. J. Agric. Res.*, **12**: 306.
(5) Barrow, N. J. (1966). *Aust. J. Exp. Agric. Anim. Husb.*, **6**: 9.
(6) Barrow, N. J. (1967). *Soil Sci.*, **104**: 242.
(7) Barrow, N. J. (1969). *Aust. J. Exp. Agric. Anim. Husb.*, **9**: 437.

(8) Blair, G. J. (1971). *J. Aust. Inst. Agric. Sci.*, **37**: 113.
(9) Chaudhry, I. A. and Cornfield, A. H. (1967). *J. Sci. Fd. Agric.*, **18**: 38.
(10) Chaudhry, I. A. and Cornfield, A. H. (1967). *J. Sci. Fd Agric.*, **18**: 82.
(11) Cooper, P. J. M. (1972). *Soil Biol. Biochem.*, **4**: 333.
(12) Cowling, D. W. and Jones, L. H. P. (1970). *Soil Sci.*, **110**: 346.
(13) Dommergues, Y. and Jacq, V. (1972). *Symposium International sur le Soufre en Agriculture Versailles*, 1970, p. 201.
(14) Fortini, S., Cardas, G. and Panella, S. (1966). *Agrochimica*, **10**: 281.
(15) Freney, J. R. (1958). *Soil Sci.*, **86**: 241.
(16) Freney, J. R. (1961). *Aust. J. Agric. Res.*, **12**: 424.
(17) Freney, J. R. (1967). In McLaren, A. D. and Peterson G. H. (eds), *Soil Biochemistry*, Marcel Dekker Inc., New York, **1**: 229.
(18) Freney, J. R., Melville, G. E. and Williams, C. H. (1969). *J. Sci. Fd Agric.*, **20**: 440.
(19) Freney, J. R., Melville, G. E. and Williams, C. H. (1971). *Soil Biol. Biochem.*, **3**: 133.
(20) Freney, J. R., Melville, G. E. and Williams, C. H. (1974) (in press).
(21) Freney, J. R. and Spencer, K. (1960). *Aust. J. Agric. Res.*, **11**: 339.
(22) Freney, J. R. and Stevenson, F. J. (1966). *Soil Sci.*, **101**: 307.
(23) Haque, I. and Walmsley, D. (1972). *Pl. Soil*, **37**: 255.
(24) Hesse, P. R. (1957). *Pl. Soil*, **9**: 86.
(25) Jenkinson, D. S. (1970). 'Rep. Rothamsted Exp. Stn for 1970', Pt 2: 113.
(26) Jones, L. H. P., Cowling, D. W. and Lockyer, D. R. (1972). *Soil Sci.*, **114**: 104.
(27) Kelly, D. P. (1972). *Symposium International sur le Soufre en Agriculture Versailles*, 1970, p. 217.
(28) Koter, M., Panak, H., Grzesuik, W. *et al.* (1965). *Roczniki Glebozn*, **15**: 37.
(29) Larson, W. E., Clapp, C. E., Pierre, W. H. and Morachan, Y. B. (1972). *Agron. J.*, **64**: 204.
(30) Lipsett, J. and Williams, C. H. (1971). *Aust. J. Exp. Agric. Anim. Husb.*, **11**: 59.
(31) Lowe, L. E. and DeLong, W. A. (1963). *Canad. J. Soil Sci.*, **43**: 151.
(32) Massoumi, A., and Cornfield, A. H. (1965). *J. Sci. Fd Agric.*, **16**: 565.
(33) Nelson, L. E. (1964). *Soil Sci.*, **97**: 300.
(34) Nicholas, D. J. (1967). *Min. Deposita*, **2**: 169.
(35) Nicolson, A. J. (1970). *Soil Sci.*, **109**: 345.
(36) Panak, K. (1966). *Agrokhimiya*, **7**: 62.
(37) Randall, P. J., Spencer K. and Freney, J. R. (1974) (in press).
(38) Scharpenseel, H. W. and Krausse, R. (1963). *Z. Pflanzenenähr Dung. Bodenk*, **101**: 11.
(39) Simon-Sylvestre, G. (1965). *Agrochimica*, **9**: 183.
(40) Simon-Sylvestre, G. (1965). *C.r. hebd. Séanc Acad. Agric. Fr.*, 426
(41) Simon-Sylvestre, G. (1967). *Agrochimica*, **10**: 60.
(42) Simon-Sylvestre, G. (1967). *C.r. hebd. Séanc Acad. Agric. Fr.*, 90.
(43) Simon-Sylvestre, G. (1969). *Ann. Agron.*, **20**: 435.
(44) Simon-Sylvestre, G. (1969). *Ann. Agron.*, **20**: 609.
(45) Simon-Sylvestre, G. (1972). *Symposium International sur le Soufre en Agriculture Versailles*, 1970, p. 181.
(46) Starkey, R. L. (1956). *Ind. Eng. Chem.*, **48**: 1429.
(47) Stewart, B. A., Porter, L. K. and Viets, F. G. Jr. (1966). *Proc. Soil Sci. Soc. Am.*, **30**: 355.
(48) Stewart, B. A., Porter, L. K. and Viets, F. G. Jr. (1966). *Proc. Soil Sci. Soc. Am.*, **30**: 453.
(49) Stewart, B. A. and Whitfield, C. J. (1965). *Proc. Soil Sci. Soc. Am.*, **29**: 752.
(50) Stotzky, G. and Norman, A. G. (1961). *Archiv. für Mikrobiologie*, **40**: 370.
(51) Syers, J. K., Adams, J. A. and Walker, T. W. (1970). *J. Soil Sci.*, **21**: 146.
(52) Tabatabai, M. A. and Bremner, J. M. (1970). *Proc. Soil Sci. Soc. Am.*, **34**: 225.
(53) Tabatabai, M. A. and Bremner, J. M. (1970). *Proc. Soil Sci. Soc. Am.*, **34**: 427.
(54) Tabatabai, M. A. and Bremner, J. M. (1972). *Agron. J.*, **64**: 40.
(55) Tabatabai, M. A. and Bremner, J. M. (1972). *Soil Sci.*, **114**: 380.

(56) Walker, T. W. (1957). *J. Br. Grassld Soc.*, **12**: 10.
(57) White, J. G. (1959). *New Zealand J. Agric. Res.*, **2**: 255.
(58) Whitehead, D. C. (1964). *Soils and Fertilizers*, **27**: 1.
(59) Williams, C. H. (1967). *Pl. Soil*, **26**: 205.
(60) Williams, C. H. (1968). *Aust. J. Soil Res.*, **6**: 131.
(61) Williams, C. H. and Donald, C. M. (1957). *Aust. J. Agric. Res.*, **8**: 179.
(62) Williams, C. H. and Lipsett, J. (1961). *Aust. J. Agric. Res.*, **12**: 612.

The Oxidation of Elemental Sulphur and Sulphides in Soil

R. G. WEIR

Elemental sulphur can be used as a source of sulphur for plants and as a means of reducing alkalinity in soils. Interest has often been shown in using elemental sulphur to increase the solubility of rock phosphate. In Australia and New Zealand considerable amounts of elemental sulphur are used in sulphur fortified superphosphate to supplement the sulphate sulphur in the original fertilizer, whilst in a number of other countries the developments towards higher analysis in fertilizers has resulted in the removal of sulphates and attempts have been made to restore this nutrient in the elemental form. Elemental sulphur has several valuable properties such as resistance to leaching which can be an advantage in certain situations, but a major factor in the increasing interest shown in it, is its high analysis. Although it was shown some fifty years ago (55), that iron sulphides can supply plant nutrient sulphur, interest has been renewed in recent years following the work of Banath (5) in 1969.

Neither elemental sulphur nor iron sulphides can be used by plants until converted to sulphates. It is therefore important to know something of the mechanism of oxidation and the factors which affect this in order to be able to appreciate the properties, potentials and limitations of these sulphur forms.

SULPHUR OXIDIZING MICRO-ORGANISMS

Elemental sulphur, sulphides and several other inorganic sulphur compounds can be oxidized in the soil by purely chemical processes but these are usually much slower and therefore of less importance than oxidation involving micro-organisms. The best known, and usually considered to be the most important, group of sulphur oxidizing micro-organisms are the autotrophic bacteria belonging to the genus *Thiobacillus*. This group has been more closely studied than the heterotrophic fungi, actinomycetes and bacteria, whose importance in the oxidation of sulphur in the soil has often been overlooked. Vitolins and Swaby (62), however, found that heterotrophic yeasts and several genera of heterotrophic and facultative autotrophic bacteria were far more numerous than the strict autotrophs, and could play an important role in many soils. Strict autotrophs were generally low in numbers, absent from over one-third of the soils and poor competitors when oxidizable sulphur was added to soils. They were, however, more efficient than the heterotrophs where conditions suited them.

Autotrophic sulphur bacteria use carbon dioxide as their only source of carbon and obtain their energy either from the oxidation of sulphur or, in the case of the photoautotrophic bacteria, from light. The latter group are aquatic and appear to be unimportant in normal aerated soils (1,62).

There are a number of species of *Thiobacilli*. Most, including some of the better known species, *T. thiooxidans*, *T. thioparus*, *T. denitrificans* and *T. ferrooxidans* are obligate autotrophs but a few are facultative autotrophs, being able to use either oxidizable sulphur or organic substances as their energy source. Although the latter group are less well known, Vitolins and Swaby (62) found that in 288 Australian soils studied one species of facultative autotroph *T. intermedius* was more numerous than all other species of *Thiobacilli* together. Most *Thiobacilli* are strictly aerobic but *T. denitrificans* is a facultative aerobe, being able to use nitrate under anaerobic conditions as an electron acceptor in place of oxygen, in the oxidation process.

The mechanism of oxidation of sulphur by autotrophic bacteria is obscure despite much investigation. Several schemes or pathways have been suggested. One by Vishniac and Santer (61) given below, proposes thiosulphate, tetrathionate, trithionate and sulphite as intermediate products, several of which have been isolated from the reaction (20,34). An alternative pathway proposed by Peck (50) involves several reactions known to occur in the reduction of sulphate.

$$S^0 \longrightarrow \underset{\text{thiosulphate}}{S_2O_3^{=}} \longrightarrow \underset{\text{tetrathionate}}{S_4O_6^{=}} \longrightarrow \underset{\text{trithionate}}{S_3O_6^{=}} \longrightarrow \underset{\text{sulphite}}{SO_3^{=}} \longrightarrow \underset{\text{sulphate}}{SO_4^{=}}$$

FACTORS AFFECTING THE OXIDATION OF ELEMENTAL SULPHUR IN SOILS

The oxidation of elemental sulphur in soils is largely a biological process. The rate of this process depends on the interaction of three types of factors namely those relating to: (a) the substrate, i.e., the sulphur; (b) the microorganisms; and (c) the environment in which the process takes place, namely, the soil.

Particle size and form of sulphur

Elemental sulphur can occur in a great variety of physical forms, but such differences within the solid forms of sulphur appear to have little effect on oxidation rate. Simon and Schollenberger (57) compared ground sulphur with sublimed forms and concluded that surface area, i.e., particle size, was more important than surface shape.

Initial sulphur oxidation rate increases as particle size is reduced. This has been frequently demonstrated in incubation studies, pot and field trials (6,16,17,26,31,32,58,64). Fox *et al.* (17) showed the relationship between specific surface of sulphur and the rate of its release as available sulphate, to be linear.

Although it is generally found that the finer the particle size the more rapid the initial rate of sulphur oxidation, there appears to be a limit beyond which improvement in sulphur recovery ceases and may even decline (16). Moreover, it is not always desirable to supply all the sulphur in its most rapidly available form. Particle size not only influences initial availability but has a potential use through the selection of differing particle sizes to give a more uniform rate of release of available sulphur over extended periods (6,26,64).

Rate and placement of sulphur

Increasing the rate of application of sulphur increases the sulphur surface available for oxidation which should result in a linear increase in the release of available sulphur. Such a relation has been shown by some workers (14,32) but in other situations (6,37) particularly in poorly buffered soils, higher rates of application can result in a marked decline in the percentage sulphur oxidized within a given period.

Band placement increases the concentration of sulphur within the oxidation zone and can reduce oxidation (22,49). This may be an effect of reduced contact between the sulphur and the micro-organism or less contact between sulphur and soil with its buffering effects. Fox *et al.* (17) overcame a depression due to banding, by mixing lime in the band suggesting the effect was, in this case, related to a build-up of harmful concentrations of acid.

Where sulphur is top-dressed, soil moisture relations could favour complete mixing of the sulphur with the soil, particularly in the field. Field work reported by Conrad (15) and Walker (63) showed top-dressing to give poor results. In a pot experiment, where surface moisture would be expected to be better maintained, no such difference between mixed and surface applications was found (18).

Microbial population and inoculation

Soils often differ markedly in the speed with which they oxidize sulphur and this is commonly attributed to differences in microbial populations. Moser and Olson (44) in comparing two soils found a high count of *T. thiooxidans* to coincide with the greater ability of one soil to oxidize sulphur. The number, types and activity of sulphur oxidizing bacteria were found to vary greatly in Australian soils (62) but *T. thiooxidans* was not prominent nor very active in these soils.

Inoculation, usually with *T. thiooxidans* (31,32) but also with heterotrophs (62) commonly show very marked improvement in the initial rate of oxidation in pot and incubation studies but such responses probably give only a short term advantage during the first month. Results from field trials usually show less benefit. The addition of sulphur and some compounds of sulphur, to the soil normally results in a rapid build-up of sulphur oxidizing micro-organisms and improved oxidizing power of the soil (6,44,62). This and the often overriding effects of environmental factors, could account for the general lack of response to inoculation seen in field trials.

Soil temperature

Microbial oxidation of sulphur will occur at temperatures as low as 4°C in the soil but the process is slow below 10°C. There is a marked increase in activity from 23°C to 30°C with further increases to 40°C (18,21,32). At temperatures above 55° to 60°C the organisms are killed (11). Optimum temperatures are not the same for all the micro-organisms responsible for sulphur oxidation but temperatures between about 25°C and 40°C are near optimum for most.

Soil moisture and aeration

Moisture and aeration in the soil are closely interrelated. Sulphur oxidizing bacteria are mostly aerobic and their activity will decline if oxygen is lacking due to waterlogging (27). Oxidation of elemental sulphur is most rapid at soil moisture levels near to field capacity (31,44). Moisture conditions which are optimum for the growth of plants also favour a high rate of sulphur oxidation.

Soil type and properties

Large differences are commonly found in the initial rate of sulphur oxidation by soils but there appears to be no relationship to soil type (28,31) or a number of soil properties such as texture (3,53,62), field moisture capacity (31), pH (31,62), initial sulphate sulphur (53) or degree of sulphur deficiency of the soil (62).

Liming may increase sulphur oxidation (9,17,24), have no effect (30,56) or inhibit oxidation (3,24). Besides its effect on pH, lime addition could affect sulphur oxidation through the reduction in soluble sulphate, aluminium, iron, manganese and other possibly toxic elements, or through increased supply of calcium or carbon dioxide.

Chopra and Kanwar (14) found that sulphur oxidation proceeded more rapidly in soils saturated with monobasic cations, sodium, potassium and hydrogen rather than divalent cations calcium and magnesium and proposed that the variable responses to lime could be due in part to the adverse effect on sulphur bacteria produced by excessive divalent cations.

Nutrient supply

The nutrient requirement of micro-organisms involved in sulphur oxidation are not fully known but micro-organisms in general require most of the nutrients which are essential for higher plants plus one or more others such as cobalt and possibly vanadium (46). The nutrient needs of the sulphur oxidizing micro-organisms appear to be met by most soils capable of growing crops. In fact they seem capable of successfully competing with higher plants, e.g., in California (41), plants were found to suffer from a temporary nitrogen shortage when sulphur was added to the soil because of stimulated activity of the sulphur oxidizing bacteria. Arif (3) found that sulphur oxidized at a reasonable rate under conditions of low phosphorus and potassium fertility but was more rapid in fertilized soil.

Sulphur incorporated into fertilizers has been reported (8,9,21) to oxidize more rapidly than sulphur added to soil alone. Grava (21) found that sulphur oxidized more rapidly with N-P fertilizers than with triple superphosphate. Bloomfield (9) found a similar effect with limed soils but the reverse in unlimed soils. While such enhanced sulphur oxidation may be due to the effects of the nutrients, nitrogen, phosphorus or calcium, they could be the result of more favourable moisture conditions around the fertilizer granule (8).

The ammonium form of nitrogen is the form required by *Thiobacteria*. Nitrates can be injurious, but sulphur bacteria are relatively tolerant of other anions, and in particular sulphates (29). This is fortunate as sulphur oxidation, particularly in alkaline soil, results in considerable increases in soluble calcium, sulphates, potassium, phosphorus, and other nutrients (37). High chloride will retard oxidation but the effect is less than that for most other biological transformations in the soil (29). High concentrations of ammonia in banded anhydrous ammonia did not depress sulphur oxidation (49).

Although heavy metals are highly toxic to most micro-organisms, some strains of sulphur oxidizing organisms show extraordinary tolerance in the oxidation of sulphur compounds from metal ores (12,47).

Organic matter

Organic matter is not essential for the activity of autotrophic sulphur bacteria but heterotrophs require a source of organic carbon. Oxidation can be rapid in soils either high or low in organic matter. Li and Caldwell (32) found that addition of ground corn cob to soil increased oxidation but Jones *et al.* (27) found little effect from the incorporation of straw in their lysimeter studies. Kittams (30) found that sucrose depressed sulphur oxidation while Vitolins and Swaby (62) found that starch had no effect on oxidation but reduced the numbers of *T. thiooxidans*.

OXIDATION OF SULPHIDES

Sulphur combines with metals to form sulphides and often polysulphides. With varying ease, most can be oxidized in well-aerated soils by chemical and/or biological processes, to produce sulphates.

Ammonium, calcium and potassium polysulphides have been used in the United States as sulphur fertilizers and as soil conditioners to acidify alkaline soils (36,60). They are water soluble and are normally applied in liquid form, either directly to the soil or in irrigation water. They are chemically unstable, forming sulphides and elemental sulphur in a finely divided state which is readily oxidized by biological means to sulphate (36).

The process of oxidation

Most investigations of the process by which iron sulphides are oxidized have been made in relation to: (a) the problems of acidity resulting from the oxidation of pyrites in coal mines; or (b) oxidation of sulphides in drained soils which were formerly flooded. Little investigation has been made of the process where sulphides are applied to the soils as fertilizers.

The oxidation of iron sulphides in soil is a more complex process than that of elemental sulphur (S^0). It is generally considered that the oxidation consists of both chemical and biological processes (25,51,52). The oxidation of both ferrous sulphide, FeS, and iron polysulphide or pyrite FeS_2 are basically similar, consisting of an initial chemical splitting of the sulphide followed by oxidation of the iron to the ferric state. This is then followed by the oxidation of elemental sulphur (S^0) to sulphuric acid and sulphates.

Temple and Delchamps (59) proposed a four step process for the oxidation of pyrite:

$$2\ FeS_2 + 2\ H_2O + 7\ O_2 \longrightarrow 2\ FeSO_4 + 2\ H_2SO_4 \quad \text{(Chemical)} \qquad (1)$$

$$4\ FeSO_4 + O_2 + 2\ H_2SO_4 \longrightarrow 2\ Fe_2(SO_4)_3 + 2\ H_2O \quad (\textit{T. ferrooxidans}) \qquad (2)$$

$$Fe_2(SO_4)_3 + FeS_2 \longrightarrow 3\ FeSO_4 + 2S \quad \text{(Chemical)} \qquad (3)$$

$$2\ S + 3\ O_2 + 2\ H_2O \longrightarrow 2\ H_2SO_4 \quad (\textit{T. thiooxidans}) \qquad (4)$$

In this scheme, pyrite is first oxidized by chemical process, to ferrous sulphate (reaction (1)). Ferrous sulphate is then oxidized to the ferric salt (reaction (2)) through the action of *T. ferrooxidans*. The ferric ions so formed are then able to chemically oxidize pyrite with the release of elemental sulphur (reaction (3)) which is subject to oxidation by *T. thiooxidans* with the formation of acidity (reaction (4)). This acidity further promotes the whole process (23,24,51).

The main difference between the oxidation of iron sulphide FeS and pyrite FeS_2 is the very much greater resistance of the latter to attack by atmospheric oxygen. Consequently, the effect of *T. thiooxidans* in producing an acid environment and *T. ferrooxidans* in the production of Fe^{3+} are of much greater significance in speeding up the first step, viz. the splitting of the iron polysulphide (23).

While the above scheme appears to apply under conditions where high acidity develops, in acid mine waters, mining dumps and in drained soils sufficiently high in sulphides to bring about pH's of 3 to 4, it may not apply where soil acidity is less extreme (10). Both *T. ferrooxidans* and *T. thiooxidans* have optimum pH ranges below 4.5 and their effectiveness in neutral to mildly acid soil conditions is therefore doubtful. In soils less acid than pH 4, pyrite oxidizes relatively slowly by chemical processes. Where iron sulphides are used as a fertilizer, the change in soil pH would be small and except for those soils which are initially very acid, conditions would not appear to favour the activity of these two *Thiobacilli*. However, the pH within the micro-environment surrounding an oxidizing sulphide particle may be considerably more acid than is measured in the bulk of the soil (25).

Bacteria in the oxidation of sulphides

The rate of oxidation of iron sulphides can be greatly enhanced by the activity of certain micro-organisms. Rudolfs (55) observed increased oxidation of pyrites following addition of a soil inoculum. Under ideal conditions the increase in the rate of oxidation of pyrites can be as much as 8 to 13 fold due to the presence of iron and sulphur oxidizing bacteria (54). Rasmussen (52) and Bloomfield (10) concluded that the oxidation of pyrite was a relatively slow process, but it was fastest in the presence of appropriate bacteria and when conditions favoured their multiplication.

In trials using iron sulphides as sources of fertilizer sulphur, Banath (5) found no benefit from the inoculation of pyrites with *T. ferrooxidans*. It was not shown whether the appropriate micro-organisms were already present in the soil or the oxidation was non-biological. Barrow (7), however, found the oxidation of pyrrhotite was reduced by the bacterial inhibitor, mercuric chloride. He also observed that the rate at which elemental sulphur and pyrrhotite were oxidized by different soils was related, suggesting the same organism was involved.

FACTORS AFFECTING THE OXIDATION OF IRON SULPHIDES

Chemical and physical form of iron sulphide

Iron sulphides can occur as the monosulphide Fe S (troilite), as polysulphides including pyrite and marcasite FeS_2 or as pyrrhotite Fe_{n-1} S_n.

Rogoff (54) reported that the crystalline form of FeS_2 affected the rate of biological oxidation. *T. thiooxidans* can directly promote the oxidation of marcasite but not pyrite (59). Marcasite, used as a fertilizer, gave more rapid production of plant available sulphate than iron pyrite (43) but the work did not determine whether this was due to mineralogical form or to the finer particle size of the marcasite used. Barrow (7) found that fine pyrrhotite oxidized at about the same rate as elemental sulphur but pyrite was much slower. Banath (5) also found pyrites very slow to oxidize. This could be overcome by using high rates of application and very fine grinding.

Instances have been reported (23) of very high resistance of some forms of pyrites to oxidation in soil. Pyrite was found in some arable soils 100 years after they had been drained. This high resistance was thought to be associated with high S:Fe ratio in the polysulphides (often in excess of 3:1), and to the formation of iron phosphates.

Particle size

As with the oxidation of elemental sulphur, the rate of oxidation of iron sulphides is increased by decreasing the particle size (5,7,54,59). A greater degree of grinding is apparently necessary to obtain a reasonable rate of oxidation with pyrite than is necessary for elemental sulphur (5,7,59) or pyrrhotite (7).

Apart from microbial population the main soil factors reported to affect oxidation of iron sulphides in the soil are, pH, phosphate status and aeration.

Soil pH and liming

Oxidation of pyrite has been reported to occur in soils containing free lime (23) but it is usually found to be much faster at low pH's than at neutral or alkaline reaction (51,52,54). Liming seems to have a greater inhibitory effect on the rate of oxidation of pyrite than it does with elemental sulphur (24,35). Although the oxidation of elemental sulphur is one of the steps in the oxidation of pyrite (59), it is not the most critical one in relation to the effect of liming. Low pH favours the oxidation of pyrite by both atmospheric oxygen and ferric ions and also promotes the biological oxidation of Fe^{2+} by *T. ferrooxidans*.

Nutrients

Although phosphate can stimulate the oxidation of elemental sulphur (7,9), its effect with iron sulphides is more often to depress oxidation (7,51). This inhibitory effect appears to be due to the formation of iron phosphates which block the oxidation and release of iron from the surface of the pyrite but the effect is not always consistent. Barrow (7) found that phosphate depressed oxidation in six out of eight soils but increased oxidation in the other two, while Banath (5) found no evidence of phosphate depressing response from pyrites. Murakami (45) found that the effect of phosphate was pH dependent, retarding the oxidation at neutral pH but having zero effect below pH 4.

Many natural sulphide ores contain appreciable amounts of heavy metals such as copper which are toxic to many micro-organisms. However, these metals do not appear to inhibit oxidation and *T. ferrooxidans* has a remarkable tolerance to quite high concentrations of copper and other heavy metals (10,12).

The addition of a small quantity of elemental sulphur was reported (55) to stimulate the oxidation of pyrite but Banath found (5) no effect from such 'priming' treatment.

Temperature, moisture and aeration

Oxidation of pyrite is most rapid under conditions which favour the activity of the *Thiobacilli*, *T. thiooxidans* and *T. ferrooxidans* (52) so it would be expected that soil conditions such as moisture status, aeration and temperature would affect the oxidation of pyrite in similar ways to their effects on the oxidation of elemental sulphur. Cooling or partly drying the soil retards oxidation (52). Murakami (45) found oxidation to be slow at low moisture, rapid above 40 per cent of maximum waterholding capacity and most rapid at 64 per cent. Pyrite will oxidize much more quickly near the surface but can be very slow in lower layers even where drainage is good (52). The rate of oxidation is controlled by the diffusion rate of oxygen into the soil and this declines rapidly with depth from the surface. In a pot trial, however, Barrow (7) found waterlogging caused no decrease in response to either pyrite or pyrrhotite although response to sulphate and elemental sulphur fertilizers were decreased.

In special environments such as waterlogged situations iron sulphides may have a place as their availability appears to be less affected than sulphate and elemental sulphur (8).

CONCLUSIONS AND SOME AGRONOMIC IMPLICATIONS

Elemental sulphur and sulphides not only provide alternative sources of sulphur to the sulphate forms but because they must be first oxidized before they are available to plants, their availability pattern is often different. A degree of control over the availability is possible through particle size selection (7,26,64), use of more resistant forms such as pyrites and by combining forms as in sulphur-fortified superphosphates.

Because elemental sulphur and sulphides become available through biological processes, some environments will be less suitable for their use. In very cold situations (48), in drier areas or where rainfall is restricted to the colder months (15,38,40), they are often less effective than sulphate forms. Seasonally, elemental sulphur is likely to release more available sulphur in summer and autumn periods (42,64), and to perform poorly, if applied in the winter (39), or where demands for early spring growth are high (13).

Advantage can be taken of the influence of climate on the oxidation rate of elemental sulphur and sulphides in reducing sulphur losses, e.g., where cold wet conditions restrict plant utilization and favour leaching of sulphate. However, this is not restricted to the effect on sulphur availability. For example, the rate of release of nitrogen from sulphur-coated urea is reduced in cold (2) or flooded (19) soils, and is greatest when growing conditions are optimal, thereby reducing losses and resulting in improved recovery in the crop (33).

REFERENCES

(1) Alexander, M. (1961). *Introduction to Soil Microbiology*. John Wiley & Sons, New York.
(2) Allen, S. E., Hunt, C. M. and Terman, G. L. (1971). *Agron. J.*, **63**: 529.
(3) Arif, A. A. (1967). PhD thesis, Auburn University, Ala. U.S.A.
(4) Attoe, O. J. and Olson, R. A. (1966). *Soil Sci.*, **101**: 317.
(5) Banath, C. L. (1969). *Aust. J. Agric. Res.*, **20**: 697.
(6) Barrow, N. J. (1971). *Aust. J. Exp. Agric. Anim. Husb.*, **11**: 211.
(7) Barrow, N. J. (1971). *Aust. J. Exp. Agric. Anim. Husb.*, **11**: 217.
(8) Beaton, J., Hubbard, W. A., Speer, R. C. and Gardiner, R. T. (1968). *Sulphur Inst. J.*, **4** (4): 4.
(9) Bloomfield, C. (1967). *Soil Sci.*, **103**: 219.
(10) Bloomfield, C. (1972). *J. Soil Sci.*, **23**:1.
(11) Breed, R. S., Murray, E. G. D. and Smith, R. (1957). *Bergey's Manual of Determinative Bacteriology*. Williams and Wilkins, London.
(12) Bryner, L. C. and Jameson, A. K. (1958). *Appl. Microbiol.*, **6**: 281.
(13) Caldwell, A. C., Seim, E. C., Rehm, G. W. and Grava, J. (1972). Univ. of Minn. Agric. Exp. Sta. Tech. Bull. No. 284.
(14) Chopra, S. L. and Kanwar, J. S. (1968). *J. Indian Soc. Soil Sci.*, **16**: 83.
(15) Conrad, J. P. (1950). *Soil Sci.*, **70**: 43.
(16) Fenster, W. E. (1965). MS thesis Univ. Wisconsin U.S.A.
(17) Fox, R. L., Atesalp, H. M., Kampbell, D. H. and Rhoades, H. F. (1964). *Proc. Soil Sci. Soc. Am.*, **28**: 406.
(18) Fox, R. L., Flowerday, A. D., Hosterman, F. W., Rhoades, H. F. and Olson, R. A. (1964). Nebraska Agric. Exp. Sta. Bull. No. 214.
(19) Giordano, P. M. and Mortvedt, J. J. (1970). *Agron. J.*, **62**: 612.
(20) Gleen, H. and Quastel, J. H. (1953). *Appl. Microbiol.*, **1**: 70.
(21) Grava, J. (1965). *Proc. Midwest Fertil. Conf.*, **17**: 49.
(22) Hassan, N. and Olson, R. A. (1966). *Proc. Soil Sci. Soc. Am.*, **30**: 284.
(23) Harmsen, G. W. (1954). *Pl. Soil*, **5**: 324.
(24) Hart, M. G. R. (1959). *Pl. Soil*, **11**: 215.
(25) Hart, M. G. R. (1962). *Pl. Soil*, **17**: 87.
(26) Jones, M. B. and Ruckman, J. E. (1969). *Agron. J.*, **61**: 936.
(27) Jones, M. B., Williams, W. A. and Martin, W. E. (1971). *Proc. Soil Sci. Soc. Am.*, **35**: 542.
(28) Kampbell, D. H. (1962). Unpublished data, Nebraska Agric. Exp. Sta. given by Attoe and Olson ref. (4).
(29) Keller, P. (1969). *Pl. Soil*, **30**: 15.
(30) Kittams, H. A. (1963). PhD thesis, University of Wisconsin.
(31) Kittams, H. A. and Attoe, O. J. (1965). *Agron. J.*, **57**: 331.
(32) Li, P. and Caldwell, A. C. (1966). *Proc. Soil Sci. Soc. Am.*, **30**: 370.
(33) Locascio, S. J. and Fiskell, J. G. A. (1971). *Sulphur Inst. J.*, **7** (1): 7.
(34) London, J. and Rittenberg, S. C. (1964). *Proc. Natl. Acad. Sci. U.S.*, **52**: 1183.
(35) MacIntire, W. H., Shaw, W. M. and Young, J. B. (1930). *Soil Sci.*, **30**: 443.

(36) McGeorge, W. T., Breazeale, E. L. and Abbott, J. L. (1956). Univ. Ariz. Agri. Exp. Sta. Tech. Bull. No. 131.
(37) McGeorge, W. T. and Greene, R. A. (1935). Univ. Ariz. Agri. Exp. Sta. Tech. Bull. No. 59.
(38) McKell, C. M. and Williams, W. A. (1965). Proc. IX Int. Grassld. Congr., **1**: 439.
(39) McLachlan, K. D. and DeMarco, D. G. (1968). *Aust. J. Exp. Agric. Anim. Husb.*, **8**: 725.
(40) McLeod, C. C. (1961). *N.Z. J. Agric.*, **102**: 225.
(41) Martin, J. P. and Ervin, J. O. (1954). *Calif. Agric.*, **8**: (1), 11.
(42) Metson, A. J. (1972). *N.Z. J. Agric. Res.*, **15**: 565.
(43) Metson, A. J., Blakemore, L. C. and Chittenden, E. T. (1971). *N.Z. J. Sci.*, **14**: 104.
(44) Moser, U. S. and Olson, R. V. (1953). *Soil. Sci.*, **76**: 251.
(45) Murakami, H. (1968). *J. Sci. Soil Manure*, Tokyo, **39**: 116.
(46) Nicholas, D. J. D. (1963). *Plant Physiology III: Inorganic Nutrition of Plants.* Academic Press, New York and London edn.
(47) Nielson, R. F. and Peterson, H. B. (1972). Utah State Univ. Agr. Exp. Sta. Bull. No. 485.
(48) Odelien, M. (1966). *Sulphur Inst. J.*, **2** (4): 14.
(49) Parr, J. F. and Papendick, R. I. (1966). *Soil Sci.*, **101**: 336.
(50) Peck, H. D. (1960). *Proc. Natl. Acad. Sci. U.S.*, **46**: 1053.
(51) Quispel, A., Harmsen, G. W. and Otzen, D. (1952). *Pl. Soil*, **4**: 43.
(52) Rasmussen, K. (1963). *Dansk Kemi*, **44**: 9.
(53) Rehm, G. W. and Caldwell, A. C. (1969). *Agron. J.*, **61**: 333.
(54) Rogoff, M. H., Silverman, M. P. and Wender, I. (1960). Am. Chem. Soc., Div. Gas Fuel Chem., Preprints. **2**: 25.
(55) Rudolfs, W. (1922). *Soil Sci.*, **14**: 135.
(56) Shedd, O. M. (1928). *Soil Sci.*, **26**: 93.
(57) Simon, R. H. and Schollenberger, C. J. (1925). *Soil Sci.*, **20**: 443.
(58) Spencer, K. (1968). Fld Stn Rec. Div. Pl. Ind., CSIRO (Aust.), **7**: 1.
(59) Temple, K. L. and Delchamps, E. W. (1953). *Appl. Microbiol.*, **1**: 255.
(60) Tisdale, S. L. and Nelson, W. L. (1966). *Soil Fertility and Fertilizers.* Macmillan Co., New York.
(61) Vishniac, W. and Santer, M. (1957). *Bact. Rev.*, **21**: 195.
(62) Vitolins, M. I. and Swaby, R. J. (1969). *Aust. J. Soil Res.*, **7**: 171.
(63) Walker, T. W. (1964). *Agrochimica*, **9**: 1.
(64) Weir, R. G., Barkus, B. and Atkinson, W. T. (1963). *Aust. J. Exp. Agric. Anim. Husb.*, **3**: 314.

Reactions of Fertilizer Sulphate in Soils

N. J. BARROW

Reactions of fertilizer sulphate may be subdivided into two categories: solution and movement into the soil of sulphate fertilizers; and reactions of the sulphate ion in the soil.

SOLUTION OF SULPHATE FERTILIZERS

Fertilizers such as potassium sulphate and ammonium sulphate are very soluble. They may have specific effects in soil, e.g., the acidifying effects of ammonium sulphate, but this is largely a property of the associated cation and will not be considered here. Of greater interest are the sparingly soluble sulphate fertilizers such as the calcium sulphates. Two forms are in widespread use. In single superphosphate, calcium sulphate is mostly present as the anhydrite (22) and therefore accounts for most of the sulphate applied to soils in Australia. However, the dihydrate, gypsum, is also used as a fertilizer.

The dissolution of a sparingly soluble salt involves two sets of processes: the first are the 'chemical' steps which control the rate of movement from the solid to the solution; the second is the transport of the dissolved material away from the surface. The relative importance of these two processes has been considered by Barton and Wilde (8). They showed that under their experimental conditions the dissolution of anhydrite was limited by the chemical steps and was slower than that for gypsum. Nevertheless, they considered that anhydrite would dissolve more quickly than gypsum in soil. This apparent conflict arises because, in soil, the rate of solution would be limited by the rate of transport of the solute away from the surface. One of the important factors controlling the rate of transport is the concentration gradient. This would be steeper for anhydrite than for gypsum because anhydrite has a higher solubility. (Barton and Wilde accept values of 0.02 mole per litre for anhydrite and 0.0012 mole per litre for gypsum (8).)

The factors in addition to concentration gradient which control the rate of transport are: the physical properties of the fluid, the velocity of fluid flow, the rate of diffusion of the solute in the soil, and the size of the granules (25). Of these five factors one can be considered as constant since the fluid considered is invariably water. However, the effects of varying the other four factors on the dissolution of calcium sulphates have been studied. Thus Williams (31) showed that the rate of solution of sulphate from superphosphate granules was slower in a calcareous soil than in an acid soil. This may be attributed to a decreased solubility of calcium sulphate due to the common ion and consequent decrease in the concentration gradient. Williams (31) also showed that when the fluid phase was moving—that is, when the samples were leached—the rate of solution was greatly increased. The effects of varying the rate of diffusion of the solute have not been directly studied but the observation (31) that rate of solution decreased as the water content of the soil decreased is at least partly due to the reduced rate of diffusion.

The factor most widely studied is the size of the granules—partly because is it readily subject to practical control and partly because it has a very large effect. Thus Williams (29) found that under simulated rainfall, only 1.3 per cent of the sulphate was removed from superphosphate particles which were 5.6 mm in diameter whereas 76.7 per cent was removed from particles which were 0.16 mm. There is thus scope for controlling the rate of solution by controlling particle size. The desirable rate of solution will be affected by conditions. If soils are sandy and rainfall intense a low rate of solution may be desirable to reduce losses by leaching (4,25,27). But if leaching losses are not important a high rate of solution may be desirable for maximum effectiveness (23).

Two main techniques have been used to vary the particle size of sulphate fertilizers. Where deposits of rock gypsum are available, the degree of fragmentation can be controlled to provide a range of particle sizes (25,27). However, this approach is not always applicable. For example, Western Australian gypsum deposits are all either 'seed' gypsum or gypsum flour (kopi) (11). The other technique is to prepare granules from finely divided material. Gypsum granules have been prepared from plaster of Paris and used in laboratory experiments (1) but not in the field. Granulation of superphosphate has been more widely studied both for granules prepared in the laboratory (29,30,31,32) and for granules prepared by commercial processes (1). All results are consistent in showing that the larger the granules the slower the solution. However, there appear to be important differences between gypsum granules and superphosphate granules. On the one hand there is evidence that dissolution of sulphate from superphosphate is slower than from gypsum of equivalent size (1). The common ion effect of the high calcium concentrations as the phosphate dissolves was suggested as the most likely explanation for this (1). On the other hand, solution of superphosphate granules also involved internal surfaces (1).

REACTIONS OF THE SULPHATE ION

When the sulphate ion has entered the soil solution, several reactions are possible. Other contributors will discuss the uptake of sulphate by plants and by micro-organisms and the reduction of sulphate. This section will therefore concentrate on adsorption of sulphate and its effects on retention and availability of sulphate.

The factors which affect the adsorption of sulphate have been indicated by Harward and Reisenauer (16) in a review which provides a summary of work prior to 1966. Since then, reactions between anions and soil colloids have been intensively studied in the University of Western Australia. This has led to some general theories on anion adsorption. Since these seem to explain many of the observations on sulphate adsorption they will be outlined here. This outline is based on (18,19,26) and especially (20). These should be referred to for a more rigorous treatment.

The surface of oxides such as Al_2O_3 and Fe_2O_3 and the weathered edges of clay particles contain ions which are not fully co-ordinated in the 'ideal' lattice. These ions complete their co-ordination shells with OH groups and water molecules. The presence of H^+ ions in the solution will favour development of a positive charge on these sites. This may be balanced by an ion such as nitrate. The nitrate ion does not approach the surface closely and does not form a specific bond with the metal ion of the oxide surface. The presence of OH^- ions in solution favours the development of a negative charge, and this is balanced in a similar way by cations in the solution. For any surface there will be a pH value at which these opposing trends balance out so that the surface bears no net charge. This is called the zero point of charge (ZPC). At pH values lower than the ZPC the surface is positively charged and above it negatively charged.

The position of the ZPC is determined by the ability of the surface to act as an acid—that is to lose an H^+ ion. Thus silica is a stronger acid than goethite (FeO.OH) or gibbsite ($Al_2O_3.3H_2O$). It has its ZPC at about pH 2 and is only positively charged at very low pH values whereas the ZPC for gibbsite and goethite is about 9.

At pH values below the ZPC, the charge on the surface also depends on the ionic strength of the solution: the greater the ionic strength the greater the positive charge.

If the solution contains only anions which do not have a specific affinity for the surface atoms, e.g., NO_3^- and Cl^-, those anions are adsorbed in proportion to their concentration in the solution. However, if an anion with a specific affinity for the surface is present, e.g., sulphate, its adsorption may be so strongly preferred as to almost completely prevent adsorption of NO_3^- or Cl^-. The adsorbed sulphate approaches the surface much more closely than the NO_3^- and effects a ligand exchange reaction with water in the co-ordination shell of the surface metal atom.

Above pH 4, sulphuric acid is almost completely dissociated to H^+ and $SO_4^=$ ions. Fully dissociated anions appear to be unable to adsorb on to negative or uncharged surfaces. The amount of adsorption therefore depends on the amount of positive charge on the surface and thus is strongly affected by pH. Anions of incompletely dissociated acids, e.g., $H_2PO_4^-$, behave somewhat differently. They do not depend on the presence of positive charge and can be adsorbed on to surfaces with a net negative charge. It appears that they dissociate at the surface and that the H^+ ions so formed react with surface OH^- to form water and this is readily displaced by the anion. The amount of adsorption is much greater than for a fully dissociated anion like sulphate. Further, phosphate ions are very effective both in displacing previously adsorbed sulphate and in reducing the capacity of the surface to adsorb sulphate.

Occurrence of adsorbing soils

From the above considerations, a soil will adsorb sulphate if it has surfaces with iron or aluminium atoms which are not fully co-ordinated and if the pH is low enough for these surfaces to bear a positive charge. Adsorption will therefore tend to be well marked on soils derived from basic rocks (7). It will also be favoured by soil forming factors which produce high iron and aluminium contents rather than high silica and which favour a low pH. Both these factors may be involved in correlations between annual rainfall and ability to adsorb sulphate. Adsorption will also be favoured in younger soils (7) because the aluminium and iron oxides are not as well organized (15) and so expose a greater surface area. Where the soil forming processes have favoured illuviation of iron and aluminium to the subsoil, adsorption may be strongly developed there (12).

A question which has interested many soil scientists is whether adsorption of sulphates is mainly on iron oxides or on aluminium oxides. This question has not proved easy to answer. Both oxides are capable of adsorbing sulphate but it is difficult, if not impossible, to extract one from the soil without extracting the other. Multiple regression analyses are also hazardous because the iron and aluminium contents of soils are highly correlated and hence it is difficult to separate individual effects statistically. Although extractable aluminium has been found to be a sufficient index of sulphate adsorption (3) this does not necessarily mean that iron is not involved. With the realization that iron oxide surfaces and aluminium oxide surfaces seem to behave very similarly in anion adsorption this question may seem less important.

Measuring adsorbed sulphate present in soil

Two approaches have been used to measure the amount of adsorbed sulphate present in soil. One makes use of the finding (Chao *et al.*, 1962) that adsorbed sulphate is in kinetic equilibrium with solution sulphate and hence can be measured by isotopic exchange. This approach was used by Hasan, Fox and Boyd (17). It may, however, overestimate adsorbed sulphate because radioactive sulphate may also exchange with sulphur in soil organic matter (14). In the second approach adsorbed sulphate is displaced from the soil by shaking it with appropriate solutions. Alkaline solutions, such as the sodium bicarbonate solution of Kilmer and Nearpass (21) are effective but results are difficult to interpret because this reagent also extracts part of the organic sulphur. This disadvantage may be less marked with the 'calcium carbonate' extraction method of Williams and Steinbergs (33). In this method soil was shaken with water plus excess calcium carbonate. The adsorbed sulphate may also be displaced by solutions containing anions which are more strongly adsorbed than sulphate. A solution of potassium dihydrogen phosphate containing 500 ppm of phosphorus was used by Ensminger (12) and has subsequently been widely used. This solution displaces sulphate effectively. However, it has the disadvantage that extracts to not always filter readily. It may also extract some sulphur from organic matter. These disadvantages appear to be overcome by the 0.01 *M* solution of calcium dihydrogen phosphate introduced by Fox, Olson and Rhoades (13) and subsequently investigated further (2). On many soils, this solution contains sufficient phosphate to displace most of the adsorbed sulphate. However, on some strongly adsorbing soils it is necessary to use a wide solution soil ratio (2) or to use repeated extractions (17).

The amount of sulphur extracted by such procedures varies widely. On soils with little adsorbing ability in climates with high rainfalls, the amount may be less than one part per million. On adsorbing soils in southern Australia the topsoil may contain a few tens of parts per million of adsorbed plus soluble sulphur. On a strongly adsorbing soil from Hawaii the topsoil contained over 1,000 ppm and the subsoil over 7,000 ppm (17).

Measurement of ability to adsorb sulphate

For soils which are able to adsorb sulphate an important question is, what proportion of the sulphate is present in the solution phase? A simple way to answer this is to shake samples of soil with a series of solutions which contain known amounts of sulphate. Adsorption is then deduced from the changes in solution concentration. This figure should then be adjusted for the sulphate initially present, as measured, for example, by a separate extraction with phosphate. The calculated value for adsorbed sulphate may then be plotted against the concentration of the sulphate remaining in solution to give an adsorption isotherm.

A problem which arises when applying this method is that the conditions of measurement affect the amount of adsorption. The amount adsorbed increases with time of shaking but at a decreasing rate. Hence results are more readily repeatable if a fairly long period of shaking is used. Temperature also affects adsorption and it was found that adsorption was more marked at 4°C than at 25°C (3). This is in the opposite direction to the reported effects of temperature on phosphate adsorption (24).

The concentration and nature of the electrolyte in the medium in which adsorption is measured also affects the amount of adsorption. The higher the valency of the cation the higher the adsorption (10). The effect of concentration has been studied for calcium and the higher the concentration the greater the adsorption (6). Similar effects occur with the adsorption of other anions. For phosphate and molybdate they can be, at least partly, explained by the assumption that the surface near the adsorbed anion is negatively charged. The effects of concentration are then seen as affecting the distribution of anions between the layer near the surface and the bulk of the solution (6). A further factor is the effect of ionic strength, and of the valency of the cation, on the charge developed by the surface (28). This factor may be more important for sulphate than for phosphate or molybdate because adsorption of sulphate is more dependent on the presence of positive charge.

The above effects present the experimenter with a choice of conditions of measurement. The choice he makes should depend on his purpose. For example if he wishes to relate diffusion of sulphate to adsorption, the conditions of measurement should be as close as possible to those existing in the soil. If, on the other hand, he is interested in assigning an index of sulphate adsorption to a range of soils an arbitrary choice of conditions is possible. Probably the most important criterion is that the results should be readily comparable with measurements made by other workers. Widely used conditions are: solutions which are 0.01 *M* with respect to calcium chloride, 24 hours of shaking, and a temperature of 25 °C.

In almost all cases the adsorption isotherms for sulphate are found to be curved so that the proportion of the sulphate in solution increases as the amount present increases. Because of the curvature, at least two co-efficients are required to specify a given isotherm fully. Nevertheless it is possible to obtain a useful single-value index by measuring the amount of adsorbed sulphate at a given solution concentration, e.g., 20 ppm (5) or 32 ppm (31). This approach is effective because the shape of the adsorption isotherms is broadly similar and hence a measurement made at one concentration is highly correlated with a measurement at another concentration. The concentration chosen therefore has little effect on the relative rating of the soils. There are soils in southern Australia which retain up to 300 ppm of adsorbed sulphate when the solution concentration is 10 ppm. If the water content of the soil were 30 per cent this would mean that only 3 μg or about 1 per cent of the sulphate would be in the solution.

Effect of adsorption on retention of sulphate by soil

Consider a band of soluble sulphate fertilizer applied to the surface of a non-adsorbing soil. Water applied will displace the band downwards. Displacement will not be simple 'piston' displacement because diffusion and hydrodynamic dispersion will tend to smear the leading and trailing edges of the band. Nevertheless, application of slightly more water than is required to replace the soil water will leach the sulphate through a column of soil.

Consider now a soil which adsorbs sulphate and, as an initial simplification, that the adsorption isotherm is linear and that 90 per cent of the sulphate is adsorbed and 10 per cent in solution. Application of water moves the sulphate through the soil much more slowly and eventually produces a distribution with a secondary peak in concentration (9). This peak would not reach the bottom of the column until about 10 times the volume of soil water has leached through the column and displacement would be almost complete when about twenty times the volume of soil water had leached through. In reality, retention would be more effective than this because the adsorption isotherm would be curved so that, as the concentration dropped, the proportion in solution would decrease. Thus, even fairly weakly developed adsorption is effective in retaining sulphate against leaching.

Over pedological time, an undisturbed soil layer will tend to reach an equilibrium in which the input of sulphur equals the outgo. The input of sulphur will depend on fertilizer usage, on mineralization of organic matter, on the sulphur content of the rainfall or the irrigation water, and in some cases, accessions from the atmosphere. The outgo will depend on the volume of water which moves through the soil layer, the rate of uptake of sulphur by plants and, especially, on the ability of the soil to adsorb sulphate. Thus adsorption is only one of the factors which determine a soil's sulphate content. There will tend to be relationships between the ability of a soil to adsorb sulphate and the amount present but the nature of the relationship will depend on other factors. This is broadly the situation described by Barrow (5) and by Hasan, Fox and Boyd (17).

The effects of rainfall on a soil's sulphur status can be complex. On the one hand, the higher the rainfall the greater the tendency for the sulphate to be leached from the soil. On the other hand, high rainfall tends to favour development of adsorbing soils and high rainfall may also be associated with on-shore winds and a high input of sulphur. These factors may balance out in such a way that annual rainfall and sulphate content are positively correlated (17).

Effect of adsorption on availability of sulphate

In a soil which does not adsorb sulphate, and provided there are no losses by leaching or reduction, all of the soluble sulphate is free to move to a plant root, both by convective flow with the soil water and by diffusion. All of the sulphate is thus potentially available. The supply of sulphur to roots penetrating the soil would be high, at first, but when all the sulphate has been removed the supply may quickly change from non-limiting to limiting. That is, there would be a 'boom and bust' economy. In contrast, in a soil which adsorbs sulphate strongly, only a portion of the sulphate is free to move. Movement of sulphate to the plant root is slower but may continue for a long period (5). Root extension may also bring more of the sulphate within accessible range of the root. The supply of sulphur thus tends to be 'rationed'. Under such conditions it is not appropriate to consider any specific amount of sulphate as available. It is more appropriate to specify the factors which control the rate of supply—the amount present and the ability to adsorb sulphate.

REFERENCES

(1) Aylmore, L. A. G., Karim, M. and Quirk, J. P. (1971). *Aust. J. Soil Res.*, **9**: 21.
(2) Barrow, N. J. (1967). *Soil Sci.*, **104**: 242.
(3) Barrow, N. J. (1967). *Soil Sci.*, **104**: 342.
(4) Barrow, N. J. (1967). *Sulphur Inst. J.*, **3**: 2.
(5) Barrow, N. J. (1969). *Soil Sci.*, **108**: 193.
(6) Barrow, N. J. (1972). *Soil Sci.*, **113**: 175.
(7) Barrow, N. J., Spencer, K. and McArthur, W. M. (1969). *Soil Sci.*, **108**: 120.
(8) Barton, A. F. M. and Wilde, N. M. (1971). *N.Z. J. Sci.*, **14**: 608.
(9) Chao, T. T., Harward, M. E. and Fang, S. C. (1962). *Proc. Soil Sci. Soc. Am.*, **26**: 27.
(10) Chao, T. T., Harward, M. E. and Fang, S. C. (1963). *Proc. Soil Sci. Soc. Am.*, **27**: 35.
(11) De La Hunty, L. E. and Low, G. H. (1958). W.A. Department of Mines. Bull. No. 6.
(12) Ensminger, L. E. (1954). *Proc. Soil Sci. Soc. Am.*, **18**: 259.
(13) Fox, R. L., Olson, R. A. and Rhoades, H. F. (1964). *Proc. Soil Sci. Soc. Am.*, **28**: 243.
(14) Freney, J. R., Melville, G. E. and Williams, C. H. (1971). *Soil Biol. Biochem.*, **3**: 133.
(15) Hamblin, A. P. and Greenland, D. J. (1972). *Aust. J. Soil Res.*, **10**: 61.
(16) Harward, M. E. and Reisenauer, H. M. (1966). *Soil Sci.*, **101**: 326.
(17) Hasan, S. M., Fox, R. L. and Boyd, C. C. (1970). *Proc. Soil Sci. Soc. Am.*, **34**: 897.
(18) Hingston, F. J., Atkinson, R. F., Posner, A. M. and Quirk, J. P. (1967). *Nature* (London), **215**: 1459.
(19) Hingston, F. J., Atkinson, R. F., Posner, A. M. and Quirk, J. P. (1968). Trans. IX Int. Congr. Soil Sci., **1**: 669.

(20) Hingston, F. J., Posner, A. M. and Quirk, J. P. (1972). *J. Soil Sci.*, **23**: 177.
(21) Kilmer, V. J. and Nearpass, D. C. (1960). *Proc. Soil Sci. Soc. Am.*, **24**: 337.
(22) Lobb, W. R. and Rothbaum, H. P. (1969). *N.Z. J. Agric. Res.*, **12**: 119.
(23) McLachlan, K. D. and De Marco, D. G. (1968). *Aust. J. Exp. Agric. Anim. Husb.*, **8**: 203.
(24) Miljadi, D., Posner, A. M. and Quirk, J. P. (1966). *J. Soil Sci.*, **17**: 238.
(25) Millington, R. J. and Powrie, J. K. (1968). Trans. IX Int. Congr. Soil Sci., **3**: 385.
(26) Mott, C. J. B. (1970). Soc. Chem. Ind. London, Monograph 1970, No. 37: 40.
(27) Powrie, J. K. (1967). *J. Aust. Agric. Sci.*, **33**: 127.
(28) van Raij, B. and Peech, M. (1972). *Proc. Soil Sci. Soc. Am.*, **36**: 587.
(29) Williams, C. H. (1969). *Aust. J. Soil Res.*, **7**: 307.
(30) Williams, C. H. (1971). *Aust. J. Soil. Res.*, **9**: 83.
(31) Williams, C. H. (1971). *Aust. J. Soil Res.*, **9**: 95.
(32) Williams, C. H. and Lipsett, J. (1969). *Aust. J. Agric. Res.*, **20**: 265.
(33) Williams, C. H. and Steinbergs, A. (1962). *Pl. Soil*, **17**: 279.

The Residual Value of a Sulphur Fertilizer

K. D. McLACHLAN

Sulphate is generally accepted as the principal sulphur form used by plants. Fertilizer is applied, either as sulphate, or in a form oxidizable to it, and as such contributes directly to the soil reservoir of available sulphur. It is the fate of the sulphate in this reservoir that is important to a consideration of residual value, since the residual value of any sulphur fertilizer will be determined by the balance between the accession of sulphate from all sources to this pool in the plant root zone, and the losses from it by physical, chemical and biological processes.

Accessions to the reservoir occur through fertilizer and pesticide applications, through the use of irrigation waters, by atmospheric accretion in rain and in gaseous form, and through mineralization of the soil organic matter. Sulphur additions in fertilizers and pesticides can be controlled and quantified with some precision. This is less true of irrigation due to variability in the S content of the waters, and applies even less to accessions in rain, gas or through the mineralization of organic matter. Variations in sulphur supplied in the atmospheric additions, as much as 150-fold in rain (45), depend on climate and proximity to such sources of S supply as industrial complexes, oceans, etc. (38,45). Additions from the soil organic matter are influenced by soil factors such as pH (46) and the C:S ratio of the organic material (6), as well as by climatic factors: moisture, temperature (46,49) and desiccation (7,16).

While the sulphur accessions are of economic significance, they mask the true residual value of a sulphur fertilizer on a given soil. There is evidence of small current additions of sulphate enhancing plant use of the residual sulphur (33). On the other hand, a similar increase in sulphate supply could lead to the net loss of sulphur in a leaching situation. Nevertheless any consideration of residual value of an S fertilizer must be aware of these accessions, and of the significant influence climatic, edaphic and geographic factors in the local environment have on them.

There appear to be two direct sources of loss of sulphur from the plant's soil reservoir. The first is through volatilization. Asher and Grundon (1) have drawn attention to losses of sulphur from intact plants, and on heating plant material. It seems reasonable that similar losses might occur directly from the soil, particularly over the summer months. There is no quantitative evidence of gaseous losses in this way. However, soil sulphur balances are difficult to obtain, and volatile losses, though generally considered to be small, may be a reason for this.

The second and more important sulphate loss is through leaching from the plant root zone. Leaching of a sulphur fertilizer depends on its solubility in water, the texture and physico-chemical properties of the soil, the rate of fertilizer application (27) and on the amount of water passing through the soil. Factors affecting the solubility include the form of sulphur used; temperature; the presence of other ions; and fertilizer particle size (37). In addition to movement in the soluble form, physical removal of the fertilizer *per se* down the soil profile in the leaching process has been reported (18). Adsorption of sulphate by soil can have a moderating influence on the leaching of the ion. It is a soil characteristic, dependent on clay content (14), mineralogy (21), pH (21,51), the presence of iron and aluminium oxides (14), the presence of other ions, particularly phosphate (14,21); and on the sulphate concentration (13). By holding the ion against the leaching processes, adsorption may account for the distribution of sulphate in the soil profile (14).

There remains the chemical and biological immobilization of sulphate from the soil reservoir. This may be only a partial and temporary loss, in that sulphate removed in this way is capable of being recycled to the reservoir again. The possible formation of 'insoluble' sulphates in soils (11,50) and biological cum chemical reduction to sulphide under anaerobic conditions (15) are examples of the former process. Neither are likely to be of much significance in agricultural soils. Biological immobilization is important and is the result of sulphate uptake by micro-organisms, plants and animals. Some of the uptake may be removed from the system in plant and animal products, directly influencing residual value. The major part, however, is returned to the soil, portion in inorganic form, but most as organic matter. Factors influencing the re-mineralization of sulphate from this source have been mentioned already. It is significant, however, that even though mineralization may occur, once applied sulphur has been used in the uptake process, its residual value falls markedly, even where the bulk of the material is returned to the soil each season (32).

AGRONOMIC FACTORS

While the balance between accessions and losses determine the residual value of a sulphur fertilizer, it can be modified under the influence of man, e.g., by the introduction of the grazing animal, cultivation, or a change in the plant species.

Sulphur essentially accumulates in the older leaves of pasture plants, and defoliation by the grazing animal could be expected to place greater stress on the residual sulphur in meeting the needs for future growth. However, ingestion of sulphur by the animal results in a large part of the sulphur intake being returned to the soil, principally in the urine, and largely in inorganic form (10). This could counteract the defoliation effect by increasing the available sulphur supply to the pasture, but under such circumstances may be unevenly redistributed over the area in the animal voidings, resulting in high concentrations in localized areas, which on certain soil types could lead to increased losses during leaching. Hilder (17), was of the opinion that leaching losses were greater under animal stocking treatments. However,

May, Till and Downes (25) have been able to account for all the sulphur applied under a grazing regime in the top 25 cm of soil. McLachlan (29) compared two stocking rates on the same soil, and showed no effect of stocking rate on sulphur requirement. In fact the yield depression with sulphur additions was greater at the higher stocking rate than the lower one. Walker (41), using the relation between N and S in pasture composition, suggested that there would be less response to sulphur in grazed pastures than mown ones. Clearly a number of possibilities are open, depending on the soil type and the climate.

Under cropping and cultivation the soil sulphur decreases (24,48). With the associated breakdown in soil organic matter, fallowing increases the supply of available nutrient, and Barrow (4) has shown that renovation increases the supply of sulphur to *Phalaris tuberosa* swards.

Change in species composition affects the residual value also. Deep rooting plants make use of sulphur at levels lower down in the soil profile (12,20). Sulphur obtained in this way can be redistributed from the soil surface again, bringing the sulphur of earlier applications into the current sulphur cycle. In grass-legume associations sulphur fertilization initially favours the development of the legume (19,31,43). With the consequent rise in the nitrogen status of the sward, pastures subsequently become grass dominant. In this state, grass makes more efficient use of the residual sulphur in producing dry matter than does the legume (33). In other cases, the differences in the ability of plants to use sulphur alter the competitive ability of the species, causing considerable ecological change in the plant community (26).

MEASUREMENTS OF RESIDUAL VALUE

In the main, the assessment of residual value of a sulphur fertilizer has been by deduction from experiments designed to study other effects of sulphur on the soil-plant-animal complex. The few direct measurements that have been made, have been done in different ways, and with two main sources of sulphur fertilizer.

The superphosphate experiments

The first group concerns superphosphate experiments in which the direct effects of sulphur were also assessable. In a summer rainfall environment, a marked reduction in pasture yield occurred when the fertilizer was omitted for one season (17). The major part of that reduction was due to sulphur deficiency. In a mediterranean type climate, sulphur deficiency occurred within four years of the last application (28). Subsequently in this environment (34), the residual value of the sulphur component of superphosphate was shown to decline with time since the last application, but less markedly than the phosphorus component. While the response to applied sulphur increased with time since the last application, there was no evidence that the amount of sulphur needed to meet this demand increased. Sulphur, at the rate of 5.5 kg/ha met the demands of pastures growing in both the least and the most responsive conditions. In contrast, the phosphorus requirement was much greater under the most responsive conditions, and as a result, an increase in the P:S ratio in fertilizers for the most responsive conditions was suggested. This evidence was contrary to the deduction (47), that with increasing applications of superphosphate, and the greater build-up of organic matter under the pastures, a greater need for sulphur relative to phosphorus would arise. The difference was probably due to the mineralization of the organic sulphur in these soils.

In New Zealand, the residual value of six years of top-dressing was measured four years after the last application was made (22). While the residual values both of phosphorus and sulphur were considerable, there was a greater decline in the residual value of the sulphur than in the residual value of the phosphorus. Nevertheless, for well top-dressed, irrigated pastures, increasing responses to increasing rates of phosphorus, but not to increasing rates of sulphur, were found. Consequently, an increase in the P:S ratio in superphosphate was recommended there also.

In Western Australia, field comparisons were made on Coolup sand (8). Sulphur deficiency was common, and even occurred within the first year of omission, on a site which had forty years of previous applications. The markedly low value of residual sulphur in this situation was attributed to leaching.

The sulphur experiments

The second group of experiments involve direct assessments of sulphur fertilizer as such. In New Zealand, Walker and Adams (43) obtained pronounced residual effects from single gypsum applications over a four-year period. Other experiments were designed to assess the relative merits of gypsum and elemental sulphur, and estimate their residual value (23,43,44). While they were done in different ways, and for different purposes, an attempt has been made to reduce them to comparable terms and to summarize the results (Table 2-5).

As far as yield data are concerned, this shows that irrespective of the form of sulphur applied, the residual value declines with time. It occurs more quickly in some areas than others, and more rapidly with one form of fertilizer than the other, though the pattern does not consistently favour a particular form of sulphur fertilizer. In the New Zealand environment, there appears to be a particular advantage in the residual value of gypsum, whereas in Australia and U.S.A., the elemental form has a higher residual value. The greater initial response to gypsum in Australia and New Zealand contrasts with the effect of elemental sulphur in California. Both these facts are consistent with least leaching occurring in New Zealand and most in California, with the Australian situation intermediate between them. In contrast with the yield data, the sulphur uptake data register a consistent decline in the value of residual sulphur with time, once maximum availability of either form has been reached.

Spencer (39) assessed the residual value of a gypsum dressing by estimating the proportion of the potential response achieved by different initial treatments in subsequent years. For legumes, the residues were only one-third as effective in the second season as in the first, and only one-seventh as effective in the third season.

TABLE 2-5

Summary of residual sulphur experiments

Country	New Zealand (43)*	Australia (44)			U.S.A. (19)		New Zealand (23)		
Rate of application kg S/ha	47	34			45		56		
Form applied**	G	G	FS	CS	G	S	G	FS	CS
Years since fertilizer applied	Increase in dry matter yield over the control treatment—kg/ha								
1	1625	2225	1813	267	1961	2141	3138	1053	392
2	2555	−210	911	852	1412	2398	2634	2679	1076
3	1782	230	560	87	538	1390	3295	3340	2589
4	1457				426	897	2578	1042	1894
	Increase in sulphur uptake over the control treatment—kg/ha								
1	7.1	5.7	3.5	0.6	3.4	3.2			
2	4.4	0.0	1.8	1.3	1.6	3.4			
3	3.4	0.3	1.0	0.7	0.6	1.1			
4	1.2				0.3	0.9			

* Article number cited in the list of references.

** G = gypsum, S = elemental sulphur, FS = fine elemental sulphur, CS = coarse elemental sulphur.

There are complications in designing residual trials because measurement over a period of time involves measurement from differing base levels of each season, and interactions between seasons and the amounts of fertilizer applied. In the trials listed in Table 2-5, a range of levels was applied at one point in time and their progress measured in subsequent years. Some confounding between time and amount occurred. McLachlan and Norman (34) also measured response at one point in time, but established responses to various levels and combinations of phosphorus and sulphur in a current application, on pastures with different fertilizer histories. This was akin to measuring current responses on an already existing Mitscherlisch type response curve, which had been established by the previous superphosphate treatments on these soils. Barrow (8) measured responses by applying a common dressing of phosphorus and sulphur and subsequently omitting either or both for one or two years. Both of these systems whilst serving their purpose have inherent disadvantages.

No system is perfect, and the closest approaches to the ideal were used by Walker and Adams (42) and McLachlan and De Marco (32). In the former, the same total application in a four-year period was applied either as an initial heavy dressing, or a biennial dressing at half that rate, or annual ones at one-quarter of that rate. This again showed the high residual value of calcium sulphate dressings in New Zealand at levels up to 45 kg sulphur/ha. At double this rate, however, annual and biennial treatments gave greater yields than the initial heavy dressing. In the second approach (32), provision was made in the design for current response curves to be established each season. The responses in that season could then be compared directly with the residual effect of the same levels applied on earlier occasions. There was replication in time as well as in space. In this design, the error associated with any change in productivity of the control plots, while waiting their turn for inclusion among the treatments, was minimal. Figure 2-4 shows that the variability in control plot yields between seasons was the least possible, and in these circumstances, comment on measurement upon a shifting base is open to the least criticism.

This experiment showed that the residual value was related to the amount applied; was limited in time; could be reduced very quickly; and was markedly dependent upon seasonal conditions. Under good rainfall conditions, with greater growth and uptake there was an immediate fall in residual value in the non-leaching situation. Under similar circumstances in a leaching situation, more would be washed down the soil profile. Because the influence of season is so strong, there is little point in establishing curves discounting the value of an initial dressing in subsequent years. It is more important to recognize that once these conditions have occurred, replacement dressings will be needed.

Figure 2-4 also shows that residual soil sulphur can produce greater yields than a current application. This was due to the change to grass dominance in this treatment, and the fact that grass made more efficient use of residual sulphur than the clover (33). The high residual value of calcium sulphate dressings in New Zealand (43) was also associated with grass dominance.

The effect of residual fertilizer upon sulphur uptake (Figure 2-5) confirms

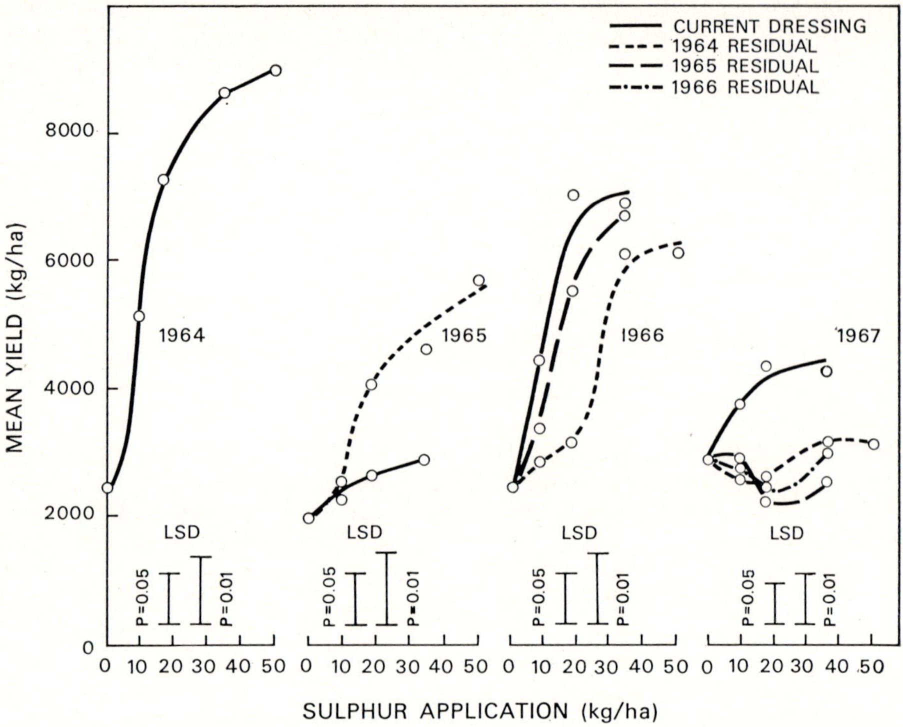

Figure 2-4 The effect of current and residual gypsum treatments on the yield of pasture in each of four years.

the more consistent pattern of decline already demonstrated when residual value is measured in these terms (19,43,44). Such data can be used to produce generalized discount curves for residual value (32) but caution is required in using them since so much depends on the seasonal conditions under which they were constructed.

FERTILIZER MANAGEMENT PRACTICES AFFECTING RESIDUAL SULPHUR

Basically, final yield is a function of adequate nutrient supply in the early stages of plant growth. As there will always be competition between plant growth, micro-organisms, and leaching for the available sulphur, there is scope for an ideal sulphur fertilizer; one with an initial rapid release of sulphur when the season opens, capable of supplying sulphur at a slower rate later on. At present, residual value can be partially influenced by fertilizer management, e.g., by changing the form of fertilizer used, the time of application or the particle size. Rate of availability, or loss, can be slowed down by using elemental sulphur, a form which has to be oxidized before it becomes available for plants. Calcium sulphate, although sparingly soluble in water, is sufficiently soluble to be readily available to plants and easily removed by leaching. Under leaching conditions (19,44) the use of elemental sulphur is

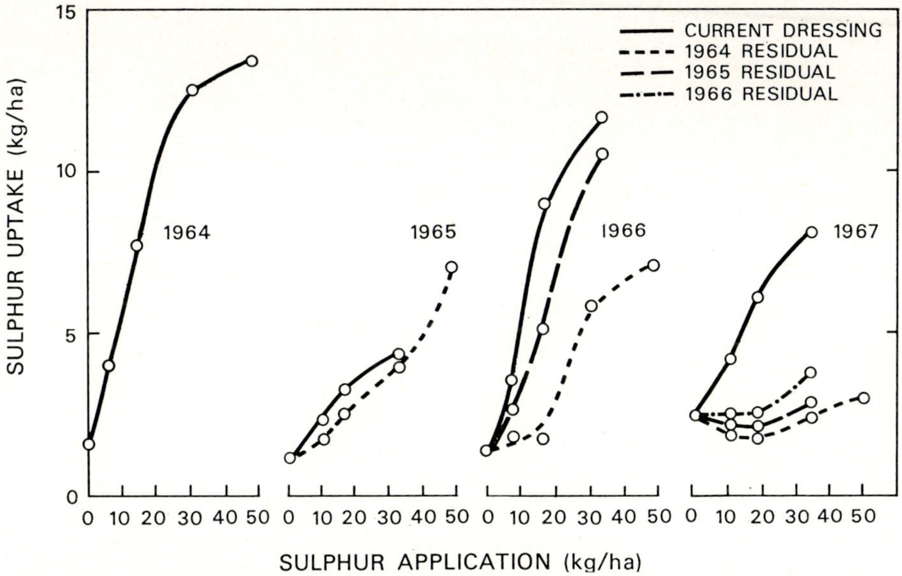

Figure 2-5 The effect of current and residual gypsum treatments on the uptake of sulphur by the total pasture species in each of four years.

favoured as it maintains responsive conditions for a longer period of time. In more arid conditions (31) calcium sulphate was more useful than elemental sulphur. Plants took advantage of any rain by immediately using the sulphate in the gypsum. There was no waiting for oxidation to occur. A combination of calcium sulphate and elemental sulphur could be useful in meeting the requirements of the ideal fertilizer. At the present time, sulphur fortified superphosphate, with its mixture of these forms, approaches closest to this ideal. Sulphides have been used as sulphur fertilizer (3,9,35) but in general the availability of the sulphur in these forms is so low that either the level of application becomes excessive or the rate of growth is too slow for economic production.

Particle size of the fertilizer affects the availability of the sulphur it contains in an inverse relationship. The availability of gypsum is influenced by the particle size (30,37,40), which in turn affects its residual value. In general, enhanced residual value is obtained at the expense of production in the earlier years. In the non-leaching situation, it has been suggested that particle size should not exceed 5 mm in diameter for efficient use (30), whereas in the leaching situation particle sizes of not less than 3 to 5 mm are needed (37). Particle size also influences the rate of oxidation of elemental sulphur (2) and can be manipulated to provide varying rates of oxidation with consequent effects on the residual value of the element (44).

Rate of application of a sulphur dressing influences its residual value. In non-leaching situations greater amounts applied led to greater residual values (Figures 2-4 and 2-5). In a leaching situation, there was a greater absolute loss of sulphur with the heavier dressings, but the percentage loss was about the same for each level of sulphur fertilizer applied (27). A greater amount could also be expected to remain in the soil from heavier dressings under these conditions.

Strategic times of application may also influence the residual value, especially in the early years. However, there is no evidence of split annual dressings being of more value than the single applications. Frequency of application also has an effect. Continued low maintenance dressings, following an initial optimum supply of sulphur, permit the reutilization of the residual sulphur in the soil, enhancing its value in maintaining production (33), probably due to the additional available sulphur source increasing the mineralization of sulphur in the organic matter (5).

To conclude, the significance of the residual value of a sulphur fertilizer lies in the fact that it determines the practical agronomic issues of when, where, how much and how often sulphur fertilizer needs to be applied. The cursory examination of factors affecting residual value indicated that climate and soil characteristics in an environment, along with the quantity of fertilizer applied, should exert the greatest influence. In practice, despite the variety in environments and experimentation reported, and the diversity of factors affecting residual value, these three things, climate, soil and amount, emerge as the principal influences on residual value. Residual value of a sulphur fertilizer is low. Added fertilizer is diverted readily to the available soil sulphur reservoir. From this point, its residual value is determined mainly by the balance attained in the interplay of factors affecting leaching, organic matter accumulation and the re-mineralization of organic sulphur. Within certain limits, the residual value can be manipulated by management practices, e.g., fertilizer form, particle size and method of application. The remedy is not simple, and generally results in forgoing some production as a consequence. The other alternative of, hopefully, smaller maintenance dressings has been effective in some pasture situations (33). Despite the present state of knowledge, it is still necessary to experiment in a particular environment. For even in the relatively simple case of accounting for leaching losses from gypsum particles in sandy soils, field results have shown that gypsum was more persistent than would have been predicted from theory (36).

REFERENCES

(1) Asher, C. J. and Grundon, N. J. (1970). *Proc. XI Int. Grassld Congr.*, p. 329.
(2) Attoe, O. J. and Olson, R. A. (1966). *Soil Sci.*, **101**: 317.
(3) Banath, C. L. (1969). *Aust. J. Agric. Res.*, **20**: 697.
(4) Barrow, N. J. (1957). *Aust. J. Agric. Res.*, **8**: 618.
(5) Barrow, N. J. (1960). *Aust. J. Agric. Res.*, **11**: 317.
(6) Barrow, N. J. (1960). *Aust. J. Agric. Res.*, **11**: 960.
(7) Barrow, N. J. (1961). *Aust. J. Agric. Res.*, **12**: 306.
(8) Barrow, N. J. (1966). *Aust. J. Exp. Agric. Anim. Husb.*, **6**: 9.
(9) Barrow, N. J. (1971). *Aust. J. Exp. Agric. Anim. Husb.*, **11**: 217.
(10) Barrow, N. J. and Lambourne, L. J. (1962). *Aust. J. Agric. Res.*, **13**: 461.
(11) Beattie, J. A. and Haldane, A. D. (1958). *Aust. J. Sci.*, **20**: 274.
(12) Blakemore, L. C., Ludecke, T. E., Leamy, M. L., *et al.* (1969). *N.Z. J. Agric. Res.*, **12**: 333.
(13) Chang, M. L. and Thomas, G. W. (1963). *Proc. Soil Sci. Soc. Am.*, **27**: 281.
(14) Ensminger, L. E. (1954). *Proc. Soil Sci. Soc. Am.*, **18**: 259.

(15) Fleming, J. F. and Alexander, L. T. (1961). *Proc. Soil Sci. Soc. Am.*, **25**: 94.
(16) Freney, J. R. (1958). *Soil Sci.*, **86**: 241.
(17) Hilder, E. J. (1954). *Aust. J. Agric. Res.*, **5**: 39.
(18) Hingston, F. J. (1959). *J. Aust. Inst. Agric. Sci.*, **25**: 209.
(19) Jones, M. B. and Ruckman, J. E. (1966). *Agron. J.*, **58**: 409.
(20) Jones, R. M. (1970). *Aust. J. Exp. Agric. Anim. Husb.*, **10**: 749.
(21) Kamprath, E. J., Nelson, W. L. and Fitts, J. W. (1956). *Proc. Soil Sci. Soc. Am.*, **20**: 463.
(22) Lobb, W. R. (1962). *N.Z. Grassld Assoc. Proc.*, **24**: 42.
(23) Ludecke, T. E. (1966). *N.Z. Grassld Assoc. Proc.*, **27**: 129.
(24) Mann, H. H. (1955). *J. Soil Sci.*, **6**: 241.
(25) May, P. F., Till, A. R. and Downes, A. M. (1968). *Aust. J. Agric. Res.*, **19**: 531.
(26) McCown, R. L. and Williams, W. A. (1968). *Ecol.*, **49**: 981.
(27) McKell, C. M. and Williams, W. A. (1960). *J. Range Mgmt*, **13**: 113.
(28) McLachlan, K. D. (1960). *Aust. J. Agric. Res.*, **11**: 513.
(29) McLachlan, K. D. (1968). *Aust. J. Exp. Agric. Anim. Husb.*, **8**: 33.
(30) McLachlan, K. D. and De Marco, D. G. (1968). *Aust. J. Exp. Agric. Anim. Husb.*, **8**: 203.
(31) McLachlan, K. D. and De Marco, D. G. (1968). *Aust. J. Exp. Agric. Anim. Husb.*, **8**: 725.
(32) McLachlan, K. D. and De Marco, D. G. (1971). *Aust. J. Exp. Agric. Anim. Husb.*, **11**: 64.
(33) McLachlan, K. D. and De Marco, D. G. (1973). *Aust. J. Exp. Agric. Anim. Husb.*, **13**: 75.
(34) McLachlan, K. D. and Norman, B. W. (1962). *Aust. J. Agric. Res.*, **13**: 836.
(35) Metson, A. J. and Collie, T. W. (1972). *N.Z. J. Agric. Res.*, **15**: 565.
(36) Millington, R. J. and Powrie, J. K. (1968). *Trans. IX Int. Congr. Soil Sci.*, **3**: 385.
(37) Powrie, J. K. (1967). *J. Aust. Inst. Agric. Sci.*, **33**: 127.
(38) Seim, E. C., Caldwell, A. C. and Rehm, G. W. (1969). *Agron. J.*, **61**: 368.
(39) Spencer, K. (1963). *Aust. J. Exp. Agric. Anim. Husb.*, **3**: 180.
(40) Spencer, K. (1968). *Fld Stn Rec. Div. Pl. Ind. CSIRO (Aust.)*, **7**: 1.
(41) Walker, T. W. (1957). *J. Brit. Grassld Soc.*, **12**: 10.
(42) Walker, T. W. (1964). *Agrochimica*, **9**: 1.
(43) Walker, T. W. and Adams, A. F. R. (1958). *Pl. Soil*, **10**: 176.
(44) Weir, R. G., Barkus, B. and Atkinson, W. T. (1963). *Aust. J. Exp. Agric. Anim. Husb.*, **3**: 314.
(45) Whitehead, D. C. (1968). *Soils Fertil.*, **27**: 1.
(46) Williams, C. H. (1967). *Pl. Soil*, **26**: 205.
(47) Williams, C. H. and Donald, C. M. (1967). *Aust. J. Agric. Res.*, **8**: 179.
(48) Williams, C. H. and Lipsett, J. (1961). *Aust. J. Agric. Res.*, **12**: 612.
(49) Williams, C. H. and Steinbergs, A. (1959). *Aust. J. Agric. Res.*, **10**: 340.
(50) Williams, C. H. and Steinbergs, A. (1962). *Pl. Soil*, **17**: 279.
(51) Williams, C. H. and Steinbergs, A. (1964). *Pl. Soil*, **21**: 50.

Sulphur Cycling in Grazed Pastures

A. R. TILL

Production from much of the world's grazing land is restricted by the limited availability of various mineral nutrients in the soil and substantial increases in animal production can often be obtained by applying fertilizer to the pasture and adjusting stock numbers to consume the herbage produced.

Efficient fertilizer use is essential and this demands a knowledge of processes and interactions in a large complex system which is influenced by uncontrolled environmental variables. The sulphur cycle in the grazed pasture is an interactive system comprised of various pools of sulphur in the soil, plants and animals, which has been studied in two ways. In one, the complete grazing situation is studied, usually in terms of fertilizer input and product output relationships, while in the other, detailed studies are made of parts of the system under controlled conditions.

In the short term, studies of the whole system are essential and provide the only reliable information for immediate use. However, the results may not apply to other sites and they give very little information on the processes and changes taking place.

The detailed studies of parts of the system, together with simulation techniques, will eventually allow a general description of the grazing system.

In this review the inputs and outputs of the system are discussed in relation to their overall influence on grazed pastures in Australia, while the pools and interactions of the recycling system are considered in more general terms.

INPUTS OF SULPHUR

Hydrogen sulphide is introduced into the atmosphere from both sea and land (17) but is rapidly converted into SO_2, SO_3, and $SO_4^=$ by a variety of processes within the atmosphere. Industrial processes and the burning of fossil fuels also introduce large amounts of sulphur into the atmosphere each year (41). The sulphur dioxide concentration in the air ranges from 0.3 $\mu g/m^3$ for unpolluted air to over 4,000 $\mu g/m^3$ for polluted air (29,43,44) but there is no evidence of a total atmospheric build-up of sulphur (3). The balance is apparently maintained by direct uptake of SO_2 by plants and soil (1,21,30,39), and removal in rain.

Sulphur in precipitation is variable, with values ranging from 0.1 kg/ha/yr in Kenya and Australia (57) to 110 kg/ha/yr in France and the U.S.A. (11, 13). This sulphur is derived from various stages of atmospheric oxidation of SO_2, and sulphate-containing aerosols from the sea. It is precipitated onto the land in different forms, in rain or dry fall, and then enters the sulphur cycle in a similar manner to fertilizer S. Spencer (pers. comm.) estimated the sulphur in precipitation over the New England area of Australia as being 1 kg/ha/yr. Using the relationship between SO_2 uptake by plants and soil and precipitation input (27) the annual input of sulphur to much of Australia's grazing land will be less than 1.5 kg/ha.

The total sulphur released by weathering is about 10 per cent of the atmospheric input to soil and plants (17) and as it is unlikely to be released near plant roots, it would not be a significant input to grazed pastures. Less than 1 per cent of the sulphur used in agriculture is as pesticides (12) and little is likely to be applied to grazed pastures in this form. Clearly, plants in most of Australia's grazing lands must derive most of their sulphur from mineralization of sulphur in the soil and from fertilizer applications.

LOSSES OF SULPHUR

Loss of volatile sulphur compounds from animals and excreta represents an extremely small proportion of the sulphur in the cycle. Sulphur-35 experiments with pasture plants grown in nutrient solution have shown loss of volatile sulphur compounds but did not allow quantitative estimates to be made (2). For plants grown in soil in pots up to 7 per cent of the total sulphur was unaccounted for and was assumed to be lost as volatile compounds (38). It is doubtful if losses of this magnitude occur from grazed pastures unless there are high inputs of sulphur.

Except in the case of severe overstocking and/or drought the losses by erosion are probably small and, using the estimated rate of removal of 40 mm soil/1,000/yr (59), it can be calculated that about 0.1 kg S/ha/yr would be lost. This figure is insignificant for a grazed pasture but it does not include removal of sulphur in organic litter. In sulphur-35 studies of grazed pasture (49,50) there was no adsorption of sulphate in forms not eventually available to plants and most of the sulphur was retained in organic form when applied at up to 10 kg/ha. Plant uptake of sulphur was correlated with the adsorbed sulphate extracted from soil by 0.01 *M* monocalcium orthophosphate (8), and these radiotracer studies in the grazing system have shown that the specific radioactivity of plant sulphur was similar to that extracted from the soil by the same reagent (51). For the soil types used in the radiotracer studies cited above it appears that adsorption did not constitute a loss of 'available' sulphur.

In lysimeter studies 10 to 90 per cent of the applied sulphur-35 was leached in periods of up to four years, depending on the development of the plant cover and whether elemental sulphur or gypsum was applied (28,34,35,58). Very little of the sulphur retained was in the adsorbed state (58). In ungrazed field trials, gradual movement of applied sulphur down a 3° sloping site has been demonstrated (10).

In grazing experiments where superphosphate had been used, Donald and Williams (14) were able to account for virtually all the P and S applied over a period of 26 years, but Hilder (24) considered applied S was lost, mainly through leaching. Leaching of radioactive fertilizer has not been observed in experiments on grazed pastures (49,51) but it must occur under some conditions, as it is dependent on the adsorptive capacity and the biological activity of the soil in relation to rainfall distribution and fertilizer application rate. In some cases the amount of fertilizer applied is so large that much of it cannot be retained by biological fixation and must almost certainly be leached out, especially if phosphate has also been applied. The interactions mentioned above are probably responsible for the very variable results obtained from the different studies of sulphur movement.

There have been many studies of the return of excreta from animals grazing small paddocks and most of these have shown increases in production. However, Sears (45) drew attention to the problems of uneven return, and for large paddocks, grazed by cattle, it was suggested that little benefit was obtained from excretal returns due to its uneven distribution and the possibility of high leaching losses (42). Hilder (25) studied faecal return in the New England area of Australia using Merino sheep and found that the camp, which comprised 3 per cent of the total area, contained about 22 per cent of the total faeces and that 10 per cent of the total area contained 34 per cent of the total faeces. The distribution of urine is not known, although this will probably contain 50 to 60 per cent of the total excreted sulphur (9,55,56).

An estimate of the likely importance of redistribution can be made by considering an improved pasture stocked at 10 sheep/ha and treated with sulphur at 25 kg/ha/yr. In such a system the returns of sulphur in excreta would be about 4 kg/ha/yr in faeces and 6 kg/ha/yr in urine. If the faecal distribution is 34 per cent on 10 per cent of the area (25), this would give a deposition (and perhaps effective loss) of 1.4 kg of faecal sulphur or 5.6 per cent of the annual sulphur application. Urinary excretion, particularly from wethers, is unlikely to follow the same pattern as faecal excretion but if it did the potential loss would be estimated as 13.6 per cent of the annual application. The corresponding figures for an unfertilized native pasture would be about 20 and 50 per cent of the estimated annual atmospheric sulphur input of 1 kg S/ha. The overall consumption and excretion by cattle at equivalent stocking intensities is probably very similar to that of sheep but the excretion pattern is likely to be more even than for Merino sheep.

Thus it would appear that, if urine and faeces had the same distribution pattern and all the sulphur in the excreta deposited on the 10 per cent of the area containing the camp site was lost, the depletion would be relatively small in fertilized pastures and probably not unreasonably large on unfertilized pastures. For the stable grazing system, the losses due to products, volatilization, erosion and adsorption can be either estimated or ignored. The major problems lie in determining the amounts lost in leaching, redistribution and immobilization, which though variable can be significant.

THE RECYCLING SYSTEM

Plants derive most of their sulphur from the soil in the form of sulphate and the replenishment of this 'available' pool is of major importance. Apart from inputs already considered, and fertilizers, the available pool is replenished by the mineralization of organic sulphur in the soil and the returns in animal excreta. Soil incubation and pot experiments have shown that mineralization and/or fixation is variable and depends on the organic matter content and fertilizer treatment (5,6,18,48,53). Interpretation of such studies in terms of the grazed pasture is frequently difficult because in most cases, only a net mineralization or production of sulphate is measured, and this in the absence of competing processes such as plant uptake and grazing. This problem is illustrated by the results (20) which showed that growing plants have a modifying effect on the relative intensities of fixation and mineralization of sulphur, the 'rhizosphere effect'. Studies in the grazed pasture following the application of sulphur-35 labelled fertilizer have shown that after 150 days only 18 per cent of the soil radioactivity is in the available pool and the remainder is apparently incorporated into organic materials (51). At this stage the total amount of sulphur in the available pool is small and mineralization from the large reserves in various organic forms is essential to maintain production. The rate of this mineralization is probably the key factor in determining production in grazed pastures (51).

Much of the soil sulphur occurs as ester sulphates which, under repeated drying and wetting, release sulphate in an available form (7,46). It has been suggested that under Australian conditions the release of sulphates by this process may be of greater importance than mineralization from other forms (7). However, in the grazed pasture it is not clear where sufficient ester sulphates would come from to maintain such a process. Excreta will supply some ester sulphates but a large proportion of the sulphur excreted is already in an available form. Radiotracer studies of soil (19), and studies on grazed pastures (Till, unpublished) have shown similar amounts of sulphur incorporated into the carbon-bonded and ester sulphate fractions, and in the field there would have to be similar breakdown rates of both fractions to avoid a steady build-up of carbon-bonded sulphur with respect to ester sulphates.

In some circumstances, where there is an inadequate supply of sulphur from the soil, plants can make up the deficit by foliar uptake of sulphur from the atmosphere (39). However, over most of Australia, atmospheric sulphur levels are so low that there is little opportunity for plants to use such a mechanism.

Animals obtain sulphur from such sources as licks, supplementary feeds, drinking water and in organic form from plants. The proportion of the above ground plant production that is consumed by animals is variable and not easy to measure but energy balance studies with sheep on fertilized improved pastures (26) have shown that 59 per cent and 93 per cent of the herbage was consumed by the animals at 10 and 30 sheep/ha respectively. The retention of ingested sulphur (5 to 30 per cent) depends on the sulphur-nitrogen-energy relationships (55), and radiotracer studies with sheep fed a ration adequate in sulphur, show that about 10 per cent of the ingested sulphur is retained (52). The amount of S excreted in the faeces is relatively constant (9,37) but urinary excretion depends on the S content of the pasture. The proportion of the total sulphur excreted in the urine ranges from 6 to 90 per cent (9,55, 56), with the proportion for a non-deficient pasture being around 50 to 60 per cent. Approximately 90 per cent of the urinary sulphur is sulphate and some of the faecal sulphur is also sulphate (55). Thus, for sheep grazing pastures, not deficient in sulphur, about 90 per cent of the ingested sulphur would be returned to the system in the excreta and of this some 50 to 60 per cent is as sulphate, mainly in the urine. The rest consists of various forms of organic sulphur.

The frequency and severity of defoliation controls the translocation of material within the plants and, depending on the relative positions of the growing points, modifies the regrowth patterns and competition between species (15). Grazing will also change the plant's ability to intercept radiation and the efficiency of photosynthesis and it has been suggested that these effects are responsible for observed increases in net primary production (54) and animal production (26). However, they are probably only partly responsible as, apart from supplementary feeding, increased animal production will only result from a greater consumption of plant material. This depends on increased incorporation into plants of nitrogen and sulphur which may become more readily available as a result of modification by the animal, to the mineralization and recycling processes. These changes are not solely due to energy penetration as grazing a grass-clover sward will cause different modifications to the soil microflora to those that occur under cutting, and increased CO_2 production and nitrogen mineralization have been observed in soil from the grazed pasture (23). Considering the strong relationships between nitrogen and sulphur, it is almost certain that grazing could increase the rate of mineralization of organic sulphur in the soil. The stocking rate modifies the proportion of the plant sulphur that is recycled via the comparatively slow plant residues and microbial mineralization processes relative to that through the rapid oxidation processes in the animal. Much of the excreted sulphate is in a form readily available to plants and, although faecal sulphur is mineralized more slowly than plant sulphur, it is considered that the overall mineralization rate will be greater than if the plant material were returned directly to the soil (7). Variations in the plant leaf area and the 'rhizosphere effect' will change the ability of the plants to utilize the available sulphur but the efficiency with which the plants compete with the soil micro-organisms for the utilization of the available nutrients is not known.

The integrated effects of some of these processes have been observed in superphosphate studies by McLachlan and Norman (36), who postulated that less rather than more fertilizer may be required to maintain pasture production at higher stocking rates. In other grazed pasture experiments it has been demonstrated that, at the same fertilizer treatment and within the range of stocking rates used, both net primary production and total animal production increased with stocking rate (26,54).

As well as nutrient benefits from excretory returns there are possibilities of increased losses of nutrients due to the influence of the animal on volatilization, leaching and deposition in campsites (40). However, these sources of loss were an integral part of the grazing experiments cited above (26,36,54) and presumably were small.

ESTIMATION OF FERTILIZER REQUIREMENTS

Measurement of the sulphur content of pasture plants is the most reliable method of detecting sulphur deficiency but such information is frequently obtained too late for corrective measures to be taken during the same growing season.

There have been many studies of sulphur uptake and dry matter production for plants grown in nutrient solutions and/or sterile soil but it is virtually impossible to use the results to predict responses in grazed pastures because of the absence of competing processes. Even where plants are grown in pots or as mono-cultures in the field, the results cannot readily be applied to the grazed situation because of lack of knowledge about their reactions to grazing and competition with other plant species.

The available sulphur is only a small proportion of the total sulphur in the surface layers of soil (59) but, because sulphate is the major source of plant sulphur, the size of the available pool has frequently been used to estimate fertilizer requirements. These tests have given very variable relationships between the soil sulphate concentration and plant production (16, 47). It was suggested that problems would arise due to the lack of attention to the replenishment of the available pool, and that the organic sulphur in the soil should be considered as 'reserve sulphur' (4). Acceptable relationships between reserve sulphur and production were obtained for some soils but the method has not received widespread use. Till and May (49) have shown that a large amount of sulphur, 'the cycling sulphur pool', is involved in production but some of the soil sulphur remains comparatively unavailable and enters the cycle only slowly. They contended that it was largely the mineralization rate of sulphur from part of the soil sulphur pool that controlled production but have not attempted to use their results to predict requirements or production.

The efficiency of use of fertilizer nutrients by crops has been studied using simple isotope dilution techniques (e.g. 'A' values, (22)) to determine the relative amounts of fertilizer and naturally occurring nutrients that are taken up by the plants. This may be satisfactory for crops but cannot be directly applied to grazed pastures because there is considerable recycling of sulphur (51) and the isotope dilution will then no longer represent a simple relationship between availability and utilization from two alternative sources.

In general it appears that although there is a very large amount of information on sulphur utilization by crops that can be used to predict fertilizer requirement and yields, there is no similar knowledge available for grazed pastures.

In conclusion, the complexity of some of the interactions between the various components of the grazed pasture system emphasizes the need for studies of processes and flow rates under conditions applying to such systems. It would appear that co-ordinated radiotracer studies on grazed pastures and parts of the system will allow interacting processes to be evaluated and a model of the sulphur cycle has been used for simulation studies (31,32). Estimates of the cycling sulphur pool and total soil sulphur (33,50,51) indicate that a considerable proportion of the soil sulphur may be unavailable in the short term ($<$ 1 year) but that some fractions of carbon-bonded sulphur must be turning over fast enough to maintain production.

REFERENCES

(1) Alway, F. J., Marsh, A. W. and Methley, W. J. (1937). *Proc. Soil Sci. Soc. Am.*, **2**: 229.
(2) Asher, C. J. and Grundon, N. J. (1970). *Proc. XI Int. Grassld Congr.*, p. 329.
(3) Badger, G. M. (1973). *Search*, **4**: 59.
(4) Bardsley, C. E. and Lancaster, J. D. (1960). *Proc. Soil Sci. Soc. Am.*, **24**: 265.
(5) Barrow, N. J. (1960). *Aust. J. Agric. Res.*, **11**: 317.
(6) Barrow, N. J. (1960). *Aust. J. Agric. Res.*, **11**: 331.
(7) Barrow, N. J. (1961). *Aust. J. Agric. Res.*, **12**: 644.
(8) Barrow, N. J. (1967). *Soil Sci.*, **104**: 242.
(9) Barrow, N. J. and Lambourne, J. L. (1962). *Aust. J. Agric. Res.*, **13**: 461.
(10) Barrow, N. J. and Spencer, K. (1959). *J. Aust. Inst. Agric. Sci.*, **25**: 208.
(11) Bertrand, G. (1935). *Ann. Agronomiques*, p. 605.
(12) Coleman, R. (1966). *Soil Sci.*, **101**: 230.
(13) Crocker, W. (1945). *Soil Sci.*, **60**: 149.
(14) Donald, C. M. and Williams, C. H. (1954). *Aust. J. Agric. Res.*, **5**: 664.
(15) Edmund, D. B. and Hoveland, C. S. (1972). *N.Z. J. Agric. Res.*, **15**: 7.
(16) Ensminger, L. E. and Freney, J. R. (1966). *Soil Sci.*, **101**: 283.
(17) Eriksson, E. (1960). *Tellus*, **12**: 63.
(18) Frederick, L. R., Starkey, R. L. and Segal, W. (1957). *Proc. Soil Sci. Soc. Am.*, **21**: 287.
(19) Freney, J. R., Melville, G. E. and Williams, C. H. (1971). *Soil Biol. Biochem.*, **3**: 133.
(20) Freney, J. R. and Spencer, K. (1960). *Aust. J. Agric. Res.*, **11**: 339.
(21) Fried, M. (1948). *Proc. Soil Sci. Soc. Am.*, **13**: 135.
(22) Fried, M. and Dean, L. A. (1952). *Soil Sci.*, **73**: 263.
(23) Hall, D. M. and Grossbard, Erna (1972). *Soil Biol. Biochem.*, **4**: 199.
(24) Hilder, E. J. (1954). *Aust. J. Agric. Res.*, **5**: 39.
(25) Hilder, E. J. (1964). *Proc. Aust. Soc. Anim. Prod.*, **5**: 211.
(26) Hutchinson, K. J. (1973). *III World Conf. Anim. Prod. Melbourne Australia.* Short communication.
(27) Johansson, O. (1959). *Kgl. Landbr. Ann.*, **25**: 57.
(28) Jones, M. B., Martin, W. E. and Williams, W. A. (1968). *Proc. Soil Sci. Soc. Am.*, **32**: 535.
(29) Junge, Chr. (1972). 'International Symposium on Sulphur in Agriculture', Versailles 1970. INRA, p. 235.
(30) Kuhn, A. and Faller, N. (1972). 'International Symposium on Sulphur in Agriculture', Versailles 1970. INRA, p. 265.
(31) May, P. F., Till, A. R. and Cumming, M. J. (1972). *J. Appl. Ecol.*, **9**: 25.
(32) May, P. F., Till, A. R. and Cumming, M. J. (1973). *J. Appl. Ecol.* (in press).
(33) May, P. F., Till, A. R. and Downes, A. M. (1968). *Aust. J. Agric. Res.*, **19**: 531.
(34) McKell, C. M. and Williams, W. A. (1960). *J. Range Mgmt*, **13**: 113.
(35) McKell, C. M. and Wilson, A. M. (1963). *Agron. J.*, **55**: 134.
(36) McLachlan, K. D. and Norman, B. W. (1966). *Aust. J. Exp. Agric. Anim. Husb.*, **6**: 22.
(37) Moir, R. J., Somers, M. and Bray, A. C. (1967-68). *Sulphur Inst. J.*, **3**: 15.
(38) Nicolson, A. J. (1970). *Soil Sci.*, **109**: 345.
(39) Olsen, R. A. (1957). *Soil Sci.*, **84**: 107.
(40) Ozanne, P. G. and Kirton, D. J. (1963). CSIRO Division Plant Industry Report. 1962-3, p. 161.
(41) Papetti, R. A. and Gilmore, F. R. (1971). *Endeavour*, **30**: 107.
(42) Peterson, R. G., Woodhouse (Jr), W. W. and Lucas, H. L. (1956). *Agron. J.*, **48**: 444.
(43) Robinson, A. J. (1971). *J. Roy. Soc. Arts*, **99**: 505.
(44) Ross, F. (1971). *New Scientist*, 13 May, p. 373.
(45) Sears, P. D. (1950). *J. Brit. Grassld Soc.*, **5**: 267.
(46) Simon-Sylvestre, G. (1972). *Int. Symp. on Sulphur in Agriculture*, Versailles 1970. INRA, p. 180.

(47) Spencer, K., Bouma, D. and Moye, D. V. (1969). *Aust. J. Exp. Agric. Anim. Husb.*, **9**: 320.
(48) Stewart, B. A., Porter, L. K. and Viets, Jr, F. G. (1966). *Proc. Soil Sci. Soc. Am.*, **30**: 355.
(49) Till, A. R. and May, P. F. (1970). *Aust. J. Agric. Res.*, **21**: 253.
(50) Till, A. R. and May, P. F. (1970). *Aust. J. Agric. Res.*, **21**: 455.
(51) Till, A. R. and May, P. F. (1971). *Aust. J. Agric. Res.*, **22**: 391.
(52) Till, A. R., May, P. F. and Cumming, M. J. (1973). *Proc. I Australian Conf. Heat and Mass Transfer*. Monash University, Victoria.
(53) Tisdale, S. L. and Nelson, W. L. (1966). *Soil Fertility and Fertilizers*, 2nd edn, Macmillan Company, New York.
(54) Vickery, P. J. (1972). *J. Appl. Ecol.*, **9**: 307.
(55) Walker, T. W. (1957). *J. Brit. Grassld Soc.*, **12**: 10.
(56) Warth, F. J. and Krishnan, T. S. (1935). *Indian J. Vet. Sci. Anim. Husb.*, **5**: 319.
(57) Williams, C. H. and Steinbergs, A. (1958). *Aust. J. Agric. Res.*, **9**: 483.
(58) Williams, W. A., McKell, C. M. and Reppert, J. N. (1964). *J. Range Mgmt*, **17**: 1.
(59) Young, A. (1969). *Nature* (London), **234**: 851.

3 Sulphur in Plant and Animal Nutrition

The Uptake and Translocation of Sulphur in Plants

D. BOUMA

The uptake of nutrient ions by the roots from the substrate, their subsequent translocation, distribution and redistribution between plant parts and organs, involves a series of steps, which may be influenced differentially by external and also internal conditions, such as the physiology and ontogeny of the plant.

This paper reviews some of the information in this area as it relates to the nutrient element sulphur.

THE MOVEMENT OF SULPHUR INTO THE ROOT

Sulphur is normally taken up as the SO_4^{2-} ion from the solution bathing the roots (31). Ions reach the 'apparent free space' (AFS) in the roots by diffusion and mass flow (13). Most of this space is in essence a continuation of the outer solution and is referred to as 'water free space' (WFS). Ions in the WFS can be removed by washing with distilled water. The AFS is generally considered to include at least the wet walls, and perhaps some of the cytoplasm of cortical cells. Most of the cytoplasm, if not all, appears to be protected against free diffusion of ions by a membrane in the outer surface of the cytoplasm, the plasmalemma (1,13). Estimates of the size of the AFS vary between 8 per cent and 15 per cent of the root volume.

Cations show a rapid entry into the other component of the AFS, the 'Donnan free space' (DFS), where they cannot be released by washing with distilled water, but can exchange with other cations in accordance with Donnan principles. Cations in the DFS are assumed to be adsorbed by fixed anions in the structure involved, mainly the cell wall. Reports on exchange-adsorption for anions are less numerous than for cations. Petersson (28), found a considerable exchange-adsorption for SO_4^{-2} in the AFS of sunflower roots. He suggested that part of the adsorption sites could be associated with the cytoplasm, thus contributing to the reversible binding of sulphate during the initial stages of uptake.

The actual transport of ions from the free space across barriers requires the expenditure of metabolic energy. Several mechanisms have been proposed to link ion transport and the supply of metabolic energy, including carriers, electron transport and others (30). Leggett and Epstein (26) established that SO_4^{2-} uptake by excised barley roots involves the attachment of sulphate to specific sites of carriers. Selenate competes with sulphate on approximately equal terms, but phosphate, nitrate and chloride do not show any measurable affinity for the binding sites. Discrimination between sulphate and selenate apparently occurs on the level of amino acid incorporation into proteins (18).

TRANSPORT OF SULPHUR ACROSS THE ROOT

Most of the lateral transport across the cortex into the xylem of the stele occurs approximately within the terminal 10 cm of the roots. In this region the cortical cells are fully vacuolated and the endodermis has formed the Casparian strip in its radial cell walls (25). The cortical cell walls, and perhaps the outer surface of the cytoplasm, are thought to be accessible to the external solution. The cytoplasm itself is considered to form a continuum, the symplasm, by means of cytoplasmic connections through the cell walls, the plasmodesmata (1,13,25). Movement of ions from cortex to stele occurs via the plasmodesmata of the endodermis, which, by virtue of its Casparian strip, provides a barrier to free diffusion of ions.

The transport and distribution of $^{35}SO_4^{2-}$ across roots of intact plants has been studied by means of micro-autoradiography. In young corn plants, under steady state conditions, sulphate transport is so rapid that there is relatively little accumulation in cortex cells, even at low transpiration rates (34). Grain density due to radioactive sulphur in the endodermis was low and approximately the same as in the cortex. Some ^{35}S accumulation was apparent in the pericycle. This increased steeply in the parenchyma and rose even further in the xylem, so that the grain density in the latter was approximately twelve times that in the endodermis and cortex. These results suggest that the endodermis 'pumped' approximately the same quantity of $^{35}SO_4^{2-}$ —ions into the xylem as it took up from the 'free space' or from the inner cortex cells. In the presence of the respiratory inhibitory azide, sulphate accumulated outside the endodermis, suggesting that the passage through the endodermis is to a large extent under metabolic control. Similar results were obtained in a study with *Vicia faba* (19). Biddulph (5), also for bean roots, found a concentration of radioactivity in the endodermis, the pericycle, the parenchyma cells at the xylem poles, and less activity in the vessels themselves. This could indicate that other stele tissues than the endodermis may be involved as a diffusion barrier to SO_4^{2-} and that the tissue around the xylem vessels may be concerned in the active transfer to the vessels.

TRANSLOCATION OF SULPHUR

Although organic sulphur compounds can be synthesized in the roots, most of the sulphur is translocated to the shoots as inorganic sulphate (7,25). In the exudate of decapitated barley plants Tolbert and Wiebe (32) could not detect any ^{35}S in organic compounds and concluded that only inorganic sulphate is translocated in the xylem sap. The roots of field peas apparently contain an active system for reducing sulphate and some of this is translocated to the shoots as methionine and to a lesser extent, as cysteine and glutathione (27). However, the contribution by the organic sulphur to the sulphur nutrition of the whole shoot system was considered to be of little significance. Out of every thousand atoms of ^{35}S leaving the root only 2 to 4 were bound to organic compounds. The results were similar for plants relying on nitrogen fixation and on nitrate nitrogen. Bollard (6) found traces of methionine in the xylem sap of a number of mono and dicotyledons and gymnosperms.

SULPHUR UPTAKE BY WHOLE PLANTS

The time course of sulphur uptake by whole plants grown with adequate sulphur supplies has been studied extensively by means of $^{35}SO_4^{2-}$. Kylin (24), for example, showed a rapid initial intake of sulphate in the roots of deseeded wheat plants. In the first 10 minutes the roots contained as much as 20 per cent of the total found in the whole plant after 60 minutes. This value increased to just below 40 per cent of the 60-minute total in 30 minutes and changed little from then on. Radioactive sulphur in the shoots, which increased slowly during the first 10 to 20 minutes, began to show a rapid increase around 30 minutes until at the end (60 minutes) they contained nearly 60 per cent of the radioactive sulphur in the plant.

Under conditions of a varying supply of sulphate, uptake depends, among other factors, on the plant's sulphur status and on the sulphate supply in the root medium. Plants of a sub-optimal sulphur status have a high internal demand (or 'sink' strength) for sulphur and show a greater sulphate uptake when the sulphate supply is restored than plants that have not been under stress. This effect is to a considerable extent independent of plant size and has been found, for example, for barley (33) and for subterranean clover (10). Sulphur uptake by subterranean clover plants, raised at 0.2, 0.5 and 4 ppm sulphur, was 303, 489 and 250 μg respectively over the first three days after being placed in complete solutions (4 ppm sulphur), while plant dry weights at the beginning were 57, 73 and 86 mg respectively (10). After the initial sulphur demands were met, sulphate uptake was more closely related to plant size and probably reflected the differences, due to the previous sulphur treatments, in root size (absorbing surface), or leaf area (transpiring surface), or both. This appears to be the usual situation under comparable external conditions and with an adequate sulphur supply. This type of response pattern is probably not confined to sulphate; it has also been found for phosphate (9).

There is some doubt regarding the relationship between the uptake and supply of sulphate, when present at levels higher than necessary for maximum growth. Some reports show sulphate uptake to be proportional to its supply until maximum growth is reached, with little or no further uptake at higher levels (4,28,29), while others show a continued uptake (17,21). There is also some evidence to suggest that this relationship varies between species and many depend on ontogenetic changes (22,23).

Although 'luxury uptake' of sulphate, in the sense of uptake in excess of the plant's requirements, does occur, most of the evidence indicates that this is not as pronounced as for other nutrient elements. Suboptimal growing conditions (light, temperature, other nutrients, etc.) may result in 'luxury uptake' characterized by an increased sulphur content of the dry matter, due to a decline in the demand. However, most of the sulphate taken up in excess of the plant's requirement remains present as sulphate (12,17,31,37).

THE DISTRIBUTION AND REDISTRIBUTION OF SULPHUR IN THE PLANT

The distribution of absorbed sulphur, as well as any redistribution during ontogeny, greatly depends on the sulphur status of the plant. In plants grown with an adequate sulphate supply, sulphate is usually distributed to the young leaves while the roots also exercise a rather powerful demand (3,24,33).

Bouma *et. al.* (12) compared the sulphur distribution in subterranean clover plants when transferred from different sulphur levels to full solutions (Table 3-1). Leaves existing on the day of transfer were subdivided into older leaves (the first 5 trifoliate leaves, cotyledons and unifoliate leaf), and the younger leaves (remaining trifoliate leaves, 4 to 7 in number depending on treatment). There was a very pronounced shift in sulphur distribution between plant parts, in response to differences in the sulphur status ('sink' strength) of the plant at transfer. In the non-deficient S_3 plants most of the sulphur translocated to the shoots was distributed to newly emerging shoots (33 per cent of the total uptake by the plant). In the S_2 plants relatively less was translocated to these shoots (26 per cent) and more to the existing shoots (36 per cent). In the S_1 plants this shift became even more apparent, most of the distribution being to the older existing leaves (31 per cent) and relatively less to the newly formed shoots. The roots retained a considerable proportion

TABLE 3-1

Total sulphur content (μg) in different parts of subterranean clover plants 32 days from sowing after 25 days in treatment solutions at three sulphur levels, and the change following transfer to full nutrient (4 ppm S)

Plant part*	S-level	Total S on day 32	Increase in total S day 32 to 35		Increase in dry weight day 32 to 35
		μg	μg	%	mg
Older shoot		99	190	(31)**	1
Younger shoot		31	70	(12)	—9
Newly-formed shoots	S_1	—	91	(15)	23
Roots	0.25 ppm	47	253	(42)	13
Whole plant		177	604		28
Older shoot		231	117	(18)	8
Younger shoot		88	115	(18)	5
Newly-formed shoots	S_2	—	166	(26)	41
Roots	1.0 ppm	95	237	(37)	23
Whole plant		414	635		77
Older shoot		368	96	(19)	9
Younger shoot		158	40	(8)	6
Newly-formed shoots	S_3	—	167	(33)	39
Roots	4.0 ppm	335	206	(40)	25
Whole plant		861	509		79

* Shoots refers to leaves + petioles. See text for explanation of subdivision of shoots (9).
** Figures in brackets express the increase for the plant part concerned as a percentage of the total for the plant.

of the sulphate taken up, but without any marked effect of the initial sulphur status.

Sulphur deficiency in subterranean clover, relying on combined nitrogen as nitrogen source, is characterized by chlorosis of the younger leaves, which spreads to older leaves and becomes more intense as the sulphur stress increases. This has also been established for beans (unpublished) and for many other crops, e.g., tomatoes (15), cotton (16) and soybeans (14). Photosynthesis during a sulphur stress declines (11), largely due to a decline in the younger chlorotic leaves, while the assimilatory capacity of the older, green leaves remains intact, at least initially. As a result, assimilates for new growth during the early stages of recovery from a sulphur stress are provided by current photosynthesis, mainly in the older leaves, and by mobilization of assimilates in the younger chlorotic leaves. Breakdown of proteins has been shown to occur under sulphur stress (2) and this may make at least some carbon assimilates available for reutilization in new growth. In Table 3-1 loss of dry matter from younger existing shoots of the S_1 plants could have accounted for at least 25 per cent of the dry matter in new shoots produced during the first 3 days of recovery from the sulphur stress. It seems likely therefore that the distribution of sulphate is largely to areas with a high sink strength, that is to the newly emerging leaves in non-deficient (S_3) plants, to the same leaves but to a lesser extent in S_2 plants, or to leaves existing at the time of removal of the stress, particularly the older ones, of more deficient plants (S_1, Table 3-1), to maintain or restore the supply of assimilates for new growth. In the younger existing, chlorotic leaves of deficient plants the main requirement for sulphur may initially be for the recovery of the photosynthetic apparatus, which may take as long as 5 days, depending on the severity of the previous sulphur stress (8).

MOBILITY OF SULPHUR IN RELATION TO REDISTRIBUTION

There have been some conflicting views on the mobility of sulphur in the plant. Biddulph *et al.* (4) concluded that the mobility in bean plants compares favourably with that for phosphorus, which is appreciable. Wood (36) considered the SO_4 ion rather immobile. He quotes results for apple trees that little SO_4 is withdrawn from leaves to storage areas, prior to leaf fall, and the absence of any migration of sulphur from storage regions to new tissues in various deciduous trees in spring. Wood and Barrien (37) found a decrease in protein-S and an increase in SO_4-S when leaves of Sudan grass were kept in the dark. No translocation of SO_4 took place to stems and roots, in contrast to amino acids and amides which increased greatly in roots and stems. It would be difficult to interpret the response patterns under sulphur stress, including the deficiency symptoms, referred to above, unless a rather restricted mobility of the element is assumed. These patterns differ markedly from those for nitrogen, where a shortage is usually characterized by proteolysis and export of nitrogen compounds from older leaves, as a rule accompanied by a yellowing and accelerated senescence of these leaves (35). How markedly the redistribution of nitrogen and phosphorus differs from that for sulphur, under deficiency conditions, is illustrated by the data in Table 3-2 ((10) and Bouma, unpublished). When subterranean clover plants raised at optimum

TABLE 3-2

Changes in nitrogen, phosphorus or sulphur content (mg) of plant parts over a period of seven days after transfer of plants raised at optimum levels of these elements to solutions without nitrogen, phosphorus or sulphur respectively

Transferred to:	Days after transfer	Old tops*	New tops*	Roots
Solution without N	Day 0	2.71	—	1.20
	Day 3	1.90	0.52	1.07
	Day 7	1.31	0.87	1.39
Net loss or gain in N	0–7	—1.40	+0.87	+0.19
Solution without P	Day 0	0.38	—	0.20
	Day 3	0.23	0.09	0.22
	Day 7	0.14	0.16	0.26
Net loss or gain in P	0–7	—0.24	+0.16	+0.06
Solution without S	Day 0	0.28	—	0.20
	Day 3	0.27	0.04	0.14
	Day 7	0.25	0.15	0.10
Net loss or gain in S	0–7	—0.03	+0.15	—0.10

* Old and new tops refer to leaves and petioles present at transfer or formed afterwards (Bouma, (10) and unpublished).

levels of nitrogen or phosphorus were transferred to solutions without nitrogen or phosphorus respectively, there were no net losses of these elements from the roots and it seems likely that nitrogen and phosphorus in new leaves and petioles formed after transfer were derived from existing leaves and petioles. By contrast, net losses of sulphur from existing leaves and petioles after transfer to solutions without sulphur were only small and at least 66 per cent of sulphur in new shoots could be accounted for by redistribution from the roots. Kylin (24) also found that most of the ^{35}S in the first leaf of deseeded wheat plants was still there as long as 24 days after transfer to solutions without sulphur. His autoradiographs indicate a small transport of ^{35}S to younger leaves, but less than in the plants in solutions containing sulphur. When subterranean clover plants, grown at different sulphur levels, were transferred to complete solutions containing ^{35}S for the first 24 hours, nearly all ^{35}S present in the leaves after the first 24 hours was still in the same leaves 6 days later (10). The redistribution of ^{35}S from older to younger leaves tended to increase with the initial sulphur status of the plant, but was much more restricted than that for ^{32}P in a similar experiment with phosphorus (9).

The evidence therefore clearly points to a restricted mobility of sulphur in the plant, particularly when compared with nitrogen and phosphorus. Sulphate taken up in excess of the plant's requirements remains present as such and some of this may be available for subsequent redistribution under stress conditions, particularly from the roots (12,36). It remains somewhat puzzling why significant quantities of sulphate remain present in leaves, even in advanced stages of sulphur stress. Cotton plants after 20 days in solutions without sulphate had virtually exhausted sulphate in stems and roots, while the leaves still contained 30 per cent of the initial sulphate (16). Shoots of wheat plants grown in solutions without sulphur for 24 days contained 30 per cent of the total sulphur as sulphate, even though the plants were obviously sulphur deficient (24). There is no obvious physiological explanation for these and similar results.

It is not clear to what extent the breakdown of organic sulphur compounds can provide sulphur for reutilization in younger plant parts. Barrien and Wood (2) concluded that the leaves of Sudan grass contain one protein rich in sulphur and another one with a lower sulphur content. Hanson *et al.* (20) found that chloroplast proteins in Sudan grass are rich in sulphur and contain approximately 70 per cent of the protein sulphur in mature leaves and concluded that during senescence, with an adequate sulphur supply to the roots, cytoplasmic protein disappears more rapidly than the chloroplast protein. Ergle and Eaton (17) also commented on the high protein sulphur content of the chloroplasts, in this case of cotton. In sulphur deficient barley a depletion of the storage sulphate occurred, followed by a breakdown in cytoplasmic protein, while chloroplast protein continued to be synthesized (33). Kylin (24) implies the reverse in his sulphur deficient wheat plants, that is a breakdown of chloroplast protein (rich in sulphur) with a continued synthesis of cytoplasmic protein. The two conflicting results are difficult to reconcile. It may be that in sulphur deficient leaves, in which chlorosis first becomes apparent, breakdown of chloroplast protein occurs, while in older, as yet unaffected leaves some synthesis of chloroplast protein continues, perhaps at the expense of cytoplasmic protein. The net effect for the plant would then depend on the balance between the two groups of leaves. The results of Bouma *et al.* (11) referred to above, in which a decline in net assimilation rate due to a sulphur stress was attributed to the chlorotic younger leaves, while the older leaves maintained their photosynthetic activity, would be in line with the view that chloroplast protein is broken down in sulphur deficient leaves. The slow recovery in rates of photosynthesis may also be related to a structural disorganization due to the breakdown of chloroplast protein.

In conclusion, sulphate sulphur content has been used extensively as an index of the sulphur status of the plant. Organic sulphur tends to remain rather stable within a fairly wide range of sulphur supply and any change in the supply is usually reflected in the sulphate content. Most of what has been said above supports this view. However, it has also been shown that there can be wide differences in the sulphur status of leaves of different ages within the same plant, at least in early stages of sulphur deficiency. The younger leaves may provide a better and earlier indication of a sulphur deficiency than a bulk plant sample 'diluted' by the older more or less intact leaves. This may be a worthwhile refinement in plant analysis and one that could be applied in many cases without serious difficulties.

REFERENCES

(1) Anderson, W. P. (1972). *Ann. Rev. Pl. Physiol.*, **23**: 51.
(2) Barrien, B. S. and Wood, J. G. (1939). *New Phytol.*, **38**: 257.
(3) Biddulph, O., Biddulph, S. F., Cory, R. and Koontz, H. (1958). *Pl. Physiol.*, **33**: 293.

(4) Biddulph, O., Cory, R. and Biddulph, S. (1956). *Pl. Physiol.*, **31**: 28.
(5) Biddulph, Susann F. (1967). *Planta*, **74**: 350.
(6) Bollard, E. G. (1957). *Aust. J. Biol. Sci.*, **10**: 286.
(7) Bollard, E. G. (1960). *Ann. Rev. Pl. Physiol.*, **11**: 141.
(8) Bouma, D. (1967). *Aust. J. Biol. Sci.*, **20**: 51.
(9) Bouma, D. (1967). *Aust. J. Biol. Sci.*, **20**: 601.
(10) Bouma, D. (1967). *Aust. J. Biol. Sci.*, **20**: 613.
(11) Bouma, D., Greenwood, E. A. N. and Dowling, E. J. (1972). *Aust. J. Biol. Sci.*, **25**: 1147.
(12) Bouma, D., Titmanis, Z. and Greenwood, E. A. N. (1972). *Aust. J. Biol. Sci.*, **25**: 1157.
(13) Brouwer, R. (1965). *Ann. Rev. Pl. Physiol.*, **16**: 241.
(14) Eaton, S. V. (1935). *Bot. Gaz.*, **97**: 68.
(15) Eaton, S. V. (1951). *Bot. Gaz.*, **112**: 300.
(16) Ergle, D. R. (1954). *Bot. Gaz.*, **115**: 225.
(17) Ergle, D. R. and Eaton, F. M. (1951). *Pl. Physiol.*, **4**: 639.
(18) Ferrari, G. and Renosto, F. (1972). *Pl. Physiol.*, **49**: 114.
(19) Gahan, P. B. and Rajan, A. K. (1965). *Exp. Cell Res.*, **38**: 204.
(20) Hanson, E. A., Barrien, B. S. and Wood, J. G. (1941). *Aust. J. Exp. Bot. Med. Sci.*, **19**:231.
(21) Jones, M. B. (1962). *Proc. Soil Sci. Soc. Am.*, **26**: 482.
(22) Jones, M. B. (1963). *Agron. J.*, **55**: 251.
(23) Jones, M. B. (1964). *Agron. J.*, **56**: 235.
(24) Kylin, A. (1953). *Physiologia Pl.*, **6**: 775.
(25) Läuchli, A. (1972). *Ann. Rev. Pl. Phys.*, **23**: 197.
(26) Leggett, J. E. and Epstein, E. (1956). *Pl. Physiol.*, **31**: 222.
(27) Pate, J. S. (1965). *Science*, **149**: 547.
(28) Petersson, S. (1966). *Physiologia Pl.*, **19**: 459.
(29) Radet, E. (1966). *Sulphur Inst. J.*, **2** (1): 11.
(30) Sutcliffe, J. S. (1962). *Mineral Salts Absorption in Plants*. Pergamon Press, Oxford.
(31) Thomas, M. D., Hendricks, R. H. and Hill, G. R. (1950). *Industr. Engin. Chem.*, **42**: 2231.
(32) Tolbert, N. E. and Wiebe, H. (1955). *Pl. Physiol.*, **30**: 499.
(33) Torii, K. and Fujiwara, A. (1967). *Tohoku J. Agric. Res.*, **17**: 89.
(34) Weigl, J. and Lüttge, U. (1962). *Planta*, **59**: 15.
(35) Williams, R. F. (1955). *Ann. Rev. Pl. Physiol.*, **6**: 25.
(36) Wood, J. G. (1942). *Chron. Bot.*, **1**: 1.
(37) Wood, J. G. and Barrien, B. S. (1939). *New Phytol.*, **38**: 265.

The Function of Sulphur in Plant Growth and Metabolism

J. W. ANDERSON

THE BIOLOGICAL DISTRIBUTION AND FUNCTION OF COMPOUNDS CONTAINING SULPHUR

Compounds ubiquitous to all cellular organisms

Sulphur is a biologically important element which occurs in compounds either in valency state —2 or valency state +6. The amino acids cysteine and methionine, which contain sulphur in the —2 valency state, are constituents of cellular proteins. The proteins of cells fulfil two functions essential to life. Catalytic proteins (enzymes) catalyze the biosynthetic and catabolic reactions required for the maintenance, growth and development of cells and organisms. Structural proteins are constituents of the various cellular membranes which act as a boundary between the organism and its external environment and delineate various subcellular organelles from the organism's own cellular environment.

Cysteine has a special role in determining the 3-dimensional shape of proteins since two cysteinyl residues can form -S-S- linkages through oxidation of their respective -SH groups. Furthermore a substantial percentage of enzymes are inhibited by sulphhydryl-group reagents such as N-ethylmaleimide and *p*-chloromercuribenzoate suggesting that the -SH group of a cysteinyl residue might be involved in the attachment of the substrate molecule to the enzyme.

In addition to their role in determining protein structure and enzyme function cysteine and methionine have other functions. Cysteine is an important metabolic intermediate in the synthesis of co-enzyme A and glutathione which are also ubiquitous to all cellular organisms. Co-enzyme A serves as a cofactor in virtually all reactions involving fatty acids whilst methionine, usually as S-adenosyl methionine, serves either directly or indirectly as a methyl group donor. In plants the methyl group of methionine is involved in the biosynthesis of lignin, pectin, chlorophyll and flavonoids. Methionine is also a substrate for the synthesis of ethylene and the vitamin thiamin.

Compounds common to plants only

Some sulphur-containing compounds, although restricted to plants, are nevertheless of general biological importance. Ferredoxin, a low molecular weight protein found in plants and some micro-organisms, is a good example. It contains an atom of iron bound via two labile sulphur atoms; this arrangement confers on the compound one of the most negative redox potentials known for a biological compound. In plants reduced ferredoxin is the first known reduced product of the light reactions of photosynthesis; it serves directly as a reducing agent for the reduction of nitrite (and possibly sulphite) and as an intermediate in the synthesis of $NADPH_2$ for the reduction of CO_2. Ferredoxin, therefore, fulfils a critical role not only in the metabolism of plants but in the whole of biology since reduced ferredoxin is the ultimate source of all biological reducing power.

Another compound found only in plants is sulpholipid (sulphoquinovose diglyceride, $S = +6$) which occurs in chloroplast membranes (14). The function of this material is not yet understood, but its ubiquity to all green photosynthetic organisms (including the blue-green algae) suggests that it is intimately involved with the structure and function of chloroplast membranes.

Some sulphate esters ($S = +6$) have been reported in plants but it is not known how widely these are distributed through the plant kingdom. The most widely known of the sulphate esters are the sulphated mucilages of the algae, e.g., κ- and λ-carrageenins of the Rhodophyta and fucoidin of the Phaeophyta (20). These mucilages are located in the continuous amorphous phase of algal cell walls, but their function is not clear. Choline sulphate has been reported in some plants; this compound is a major constituent in filamentous fungi in which it serves as a storage form of sulphur (22).

Some secondary plant products of restricted distribution and unknown function

A wide variety of sulphur-containing compounds have been described in plants. Some sulphur-containing compounds have been found in selected plant species although a further search may show them to be more widely distributed throughout the plant kingdom. Some compounds (e.g., S-methylcysteine and/or its sulphoxide derivative) have been reported in several unrelated plant species. In some instances, however, various groups of compounds appear to have a restricted distribution in a family or a closely related group of families. This is instanced by the wide range of isothiocyanates (mustard oils) in the Crucifereae (16) and the S-alkylcysteine sulphoxides in some members of the Liliaceae (31), where the relevant groups of compounds have proved to have some taxonomic value. In some species up to 90 per cent of the total sulphur content of the plant may be contained in the so-called secondary plant products. The function of these compounds is not understood, but the apparent limited distribution of most of these compounds suggests that they are unlikely to fulfil important metabolic roles. Some of these compounds have weak bacteriacidal properties (31) and it has been suggested that this property might be functionally significant in pathogenesis.

Some of the secondary sulphur-containing compounds have some significance in agriculture because of their capacity either directly or indirectly to impart vile odours, toxic properties, pleasant flavours and tantalizing odours to human and animal foodstuffs. Many of the chemical principles responsible for these characteristics are absent from undamaged plants, but may appear as the result of enzymic or non-enzymic activity following maceration of the tissue. Some examples of sulphur-containing compounds conferring desirable or undesirable odours, etc., are listed in Table 3-3; some of the compounds listed are derived from plants grown for human consumption.

TABLE 3-3
Some examples of compounds and their precursors which confer desirable or undesirable flavours and odours to plant stuffs

Plant species	Principle	Precursor	Properties of the principle
Algae	Dimethyl sulphide	Dimethyl priothetin	Vile stench of rotting seaweeds
Allium cepa (onion)	Propenyl sulphenic acid	S-Propenyl cysteine sulphoxide	Lachrymatory factor of onions
	Mercaptopropane	S-Propyl cysteine	Contributes to the aroma of onions
Asparagus	Dimercaptoisobutyric acid	?	Stench associated with urine after ingesting Asparagus
Cruciferae	Dimethyl sulphide Dimethyl thiosulphinate	} Methylcysteine sulphoxide	Odour of cabbages, cauliflower, broccoli, etc.
	Various isothiocyanates	Corresponding glucosides	Pungent biting flavour of cruciferous vegetables
	Thiocyanate	Various isothiocyanates	Toxic principle of certain cruciferous pastures
Mimosaceae	Dimercaptomethane	Djenkolic acid	Vile stench of some seeds of the Mimosaceae

SULPHUR METABOLISM IN PLANTS

Biological importance

The metabolism of cysteine and methionine is strikingly different in plants and animals. Plants utilize inorganic sulphate as the sole source of sulphur required for the synthesis of cysteine and methionine; this process involves the reduction of sulphate to sulphide prior to incorporation of sulphur into organic compounds. Animals on the other hand are unable to reduce sulphate in significant quantities and are unable to synthesize methionine from sulphide although animals can execute the reverse reaction. Animals are, therefore, dependent upon a dietary source of methionine which is supplied either directly or indirectly by plants and micro-organisms (Figure 3-1).

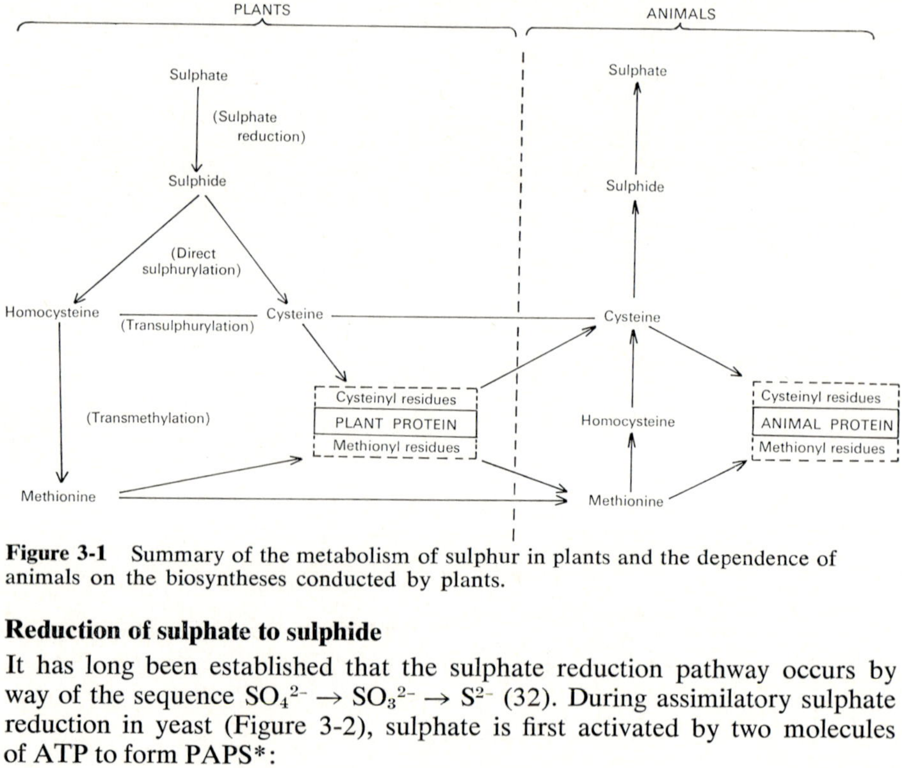

Figure 3-1 Summary of the metabolism of sulphur in plants and the dependence of animals on the biosyntheses conducted by plants.

Reduction of sulphate to sulphide

It has long been established that the sulphate reduction pathway occurs by way of the sequence $SO_4^{2-} \rightarrow SO_3^{2-} \rightarrow S^{2-}$ (32). During assimilatory sulphate reduction in yeast (Figure 3-2), sulphate is first activated by two molecules of ATP to form PAPS*:

$$1.\ ATP + SO_4^{2-} \xrightleftharpoons{\text{ATP sulphurylase}} APS + PP_i\ (\Delta G = +46\ kJ/mol);$$
$$2.\ ATP + APS \xrightleftharpoons{\text{APS kinase}} PAPS + ADP\ (\Delta G = -21\ kJ/mol).$$

PAPS serves as the activated form of sulphate which is reduced to sulphite in a reaction catalyzed by an ensyme complex called PAPS reductase (22).

It is now well established that plants contain ATP sulphurylase activity (3,24). The free energy change of the reaction is extremely unfavourable for

* Abbreviations: APS, adenosine 5′-sulphatophosphate; PAPS, adenosine 3′-phosphate 5′-sulphatophosphate; 3′-AMP, adenosine 3′-monophosphate; ΔG = free energy change.

the synthesis of APS but APS synthesis can be demonstrated if pyrophosphatase is added to hydrolyse inorganic pyrophosphate, PP_i (24). The synthesis of PAPS has been demonstrated in the unicellular alga *Chlorella* (15), but there is considerable doubt as to whether PAPS is synthesized by higher plants. Mercer and Thomas (19) reported that chloroplast extracts synthesized PAPS, but the product was not characterized. Most other authors, however, have been unable to detect the synthesis of PAPS in plants (2,3,8) and this has led to speculation that APS and not PAPS might be the activated form of sulphate in plants as is the case in the dissimilatory sulphate reducing bacteria (Figure 3-2).

The question of whether PAPS is synthesized by plants has been reinvestigated with special reference to 3′-nucleotidase activity (5) which is present in high activity in plants. If PAPS is synthesized by plants, the 3′-phosphate of PAPS could be susceptible to hydrolysis by 3′-nucleotidase thereby forming APS and explaining why PAPS is not detected. It has previously been shown that significant synthesis of PAPS in the bacterium *Salmonella pullorum* could only be demonstrated following addition of large quantities of 3′-AMP, the

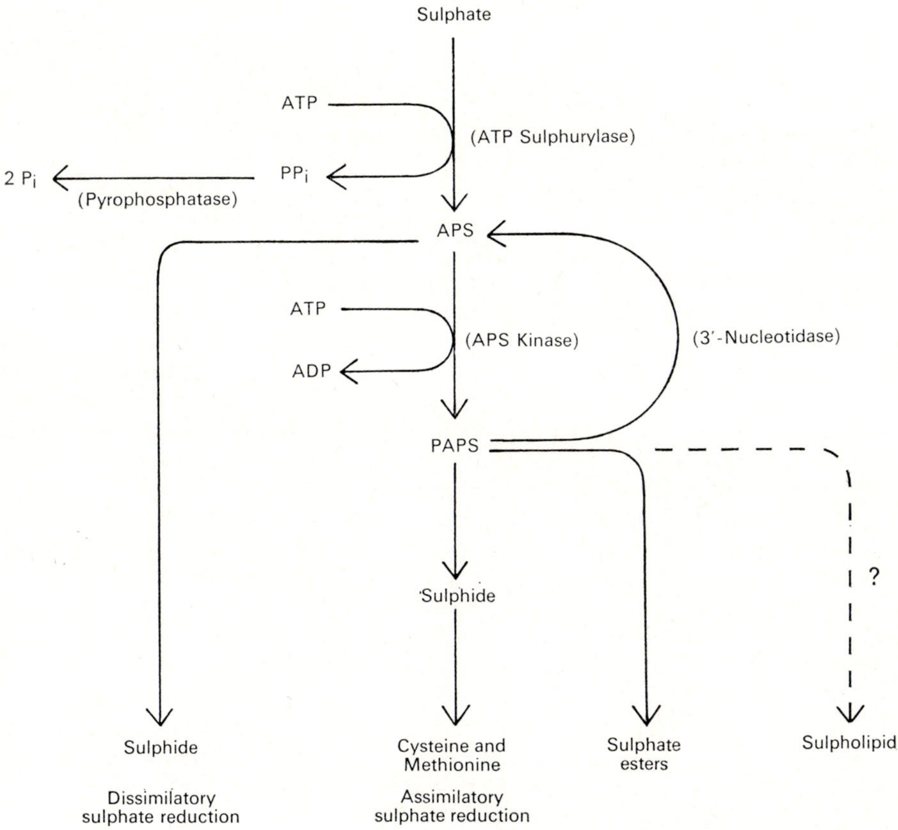

Figure 3-2 The role of APS and PAPS as intermediates in the metabolism of inorganic sulphate.

preferred substrate of 3′-nucleotidase (17). Crude extracts of spinach, in the presence of 3′-AMP, catalyze the synthesis of PAPS; PAPS was not synthesized in the absence of 3′-AMP (5). The PAPS synthesized by the spinach leaf extract was characterized by chromatography, electrophoresis, infra-red spectrophotometry and by specific enzyme catalyzed reactions. Isolated chloroplasts synthesize PAPS only in the presence of 3′-AMP although 3′-nucleotidase is absent from isolated chloroplasts suggesting that the effect of 3′-AMP on the synthesis of PAPS is not due to its action as an alternative substrate of 3′-nucleotidase. It has been suggested that in isolated chloroplasts 3′-AMP acts as an allosteric modifier of APS kinase (5).

The synthesis of PAPS does not in itself imply that PAPS is an intermediate in sulphate reduction; animal tissues for instance do not reduce sulphate but, nevertheless, synthesize PAPS which is used for the synthesis of sulphate esters (Figure 3-2) in reactions catalyzed by sulphotransferases (22). However, the incorporation of [^{35}S] sulphate into cysteine (the first detectable product of the sulphate reduction pathway) by isolated chloroplasts is enhanced six-fold by the addition of 3′-AMP (5). Since 3′-AMP is also required for the synthesis of PAPS this suggests that PAPS is an intermediate in the reduction of sulphate in higher plants (5). Although PAPS has been demonstrated in *Chlorella* (15) there is some evidence that APS is the active form of sulphate in this organism (23,30).

Little is known about the reduction of PAPS in higher plants. In yeast and *E. coli*, $NADPH_2$ is the most active electron donor. The PAPS reductase of yeast has been resolved into at least three proteins referred to as enzymes A and B and fraction C (22). Enzyme A is believed to be a flavoprotein which is reduced by $NADPH_2$. Reduced enzyme A effects the reduction of fraction C (a low molecular weight protein). Enzyme B catalyzes the reduction of the sulphate moiety of PAPS to sulphite; reduced fraction C serves as the immediate electron donor. A low molecular weight protein with properties similar to fraction C of yeast has been described in spinach chloroplasts (2). The main site of sulphate reduction in plants is thought to be the chloroplast. If the reduction of PAPS in chloroplasts is similar to the system described in yeast, then the photosynthetic products $NADPH_2$ and/or reduced ferredoxin would be the most likely electron donors. Few details of the APS reductase of *Chlorella* are available. Schmidt (23) has described an APS-specific sulphotransferase which catalyzes the transfer of the sulphate moiety to various thiols; he suggested that an unidentified endogenous thiol acted as the acceptor *in vivo* and that the bound form was the substrate for reduction.

In yeast it has been established that free sulphite is not released during sulphate reduction but sulphite remains bound to a protein though it can be exchanged if a large amount of carrier sulphite is added (22). In yeast it has been suggested that the binding protein is fraction C.

The reduction of bound sulphite is catalyzed by the enzyme complex sulphite reductase; the reduction is unusual in that it involves a 6-electron reduction of sulphite to sulphide. The enzyme has been described in spinach (1,18, 27). Unpurified enzyme catalyzes the reduction of sulphite when either reduced ferredoxin or reduced methyl viologen (an artificial reducing agent) are used as electron donors (27). Purified spinach enzyme contains iron and has some of the properties of a hemoprotein, but unlike the sulphite reductase of yeast does not contain flavin and will only use reduced methyl viologen as the electron donor (1). Asada *et al.* (1) postulated that the more specific electron donor specificity of the purified spinach enzyme was due to separation of sulphite reductase activity from the intramolecular-electron transport system of the sulphite reductase complex. Since sulphite reductase is localized in chloroplasts (18) and reduced ferredoxin serves as an electron donor of unpurified sulphite reductase, then presumably reduced ferredoxin acts as the physiological electron donor.

Biosynthesis of cysteine

The enzyme serine sulphhydrase has been described in plants (4,7). In the presence of pyridoxal phosphate the enzyme catalyzes the reaction:

3. Serine + $H_2S \rightleftharpoons$ Cysteine + H_2O.

This enzyme is associated with chloroplasts (4), the principal site of sulphide synthesis and it has long been assumed that the reaction catalyzed by serine sulphhydrase is the mechanism for the assimilation of inorganic sulphur into organic compounds (32).

Recently, an alternative pathway has been described for the biosynthesis of cysteine in plants involving the following reactions:

4. Serine + acetyl-CoA $\xrightleftharpoons{\text{Serine transacetylase}}$ O-Acetylserine + CoA;
5. O-Acetylserine + H_2S $\xrightleftharpoons{\text{O-acetylserine sulphhydrase}}$ Cysteine + Acetate.

There has been no clear demonstration that the second mechanism constitutes the major pathway for cysteine synthesis, but the following evidence suggests that this is likely: (a) the formation of cysteine by crude extracts of turnip leaf tissue was ten-fold greater with O-acetylserine than with serine (11,29); (b) O-acetylserine is an activated form of serine. It could be argued that since the free energy change of reaction (4) would be more negative than reaction (3) there would be a greater yield of cysteine at chemical equilibrium. The enzymes catalyzing reactions (4) and (5) in plants have been characterized (26). These enzymes are absent from animals though animals contain serine sulphhydrase activity implying that animals are unable to synthesize cysteine in large amounts.

Biosynthesis of methionine

The synthesis of methionine by plants and micro-organisms assumes important biological significance since animals are unable to synthesize this essential amino acid. Two pathways for the synthesis of methionine have been described in plants (28), but the relative contributions of the two pathways to the gross synthesis of methionine is not known (13). The pathways differ in the reactions leading to the synthesis of homocysteine, the key intermediate in methionine biosynthesis (Figure 3-3). One pathway (transulphurylation) involves the formation of cystathionine from cysteine and activated homoserine (probably O-acetylhomoserine in plants); cystathionine is then hydrolyzed by β-cystathionase to yield homocysteine, pyruvate and NH_3 (12). An important feature of this pathway is that it involves an activated form of homoserine which will tend to favour the synthesis of cystathionine (9); this marks an important difference in the metabolism of methionine in plants and animals (Figure 3-3).

The alternative pathway for the synthesis of homocysteine includes a direct sulphurylation of activated homoserine (probably O-acetylhomoserine) using sulphide produced by the sulphate reduction pathway as the sulphurylating agent (10,11). This reaction is analogous to reaction (5) and the same bioenergetic principles apply (Figure 3-3).

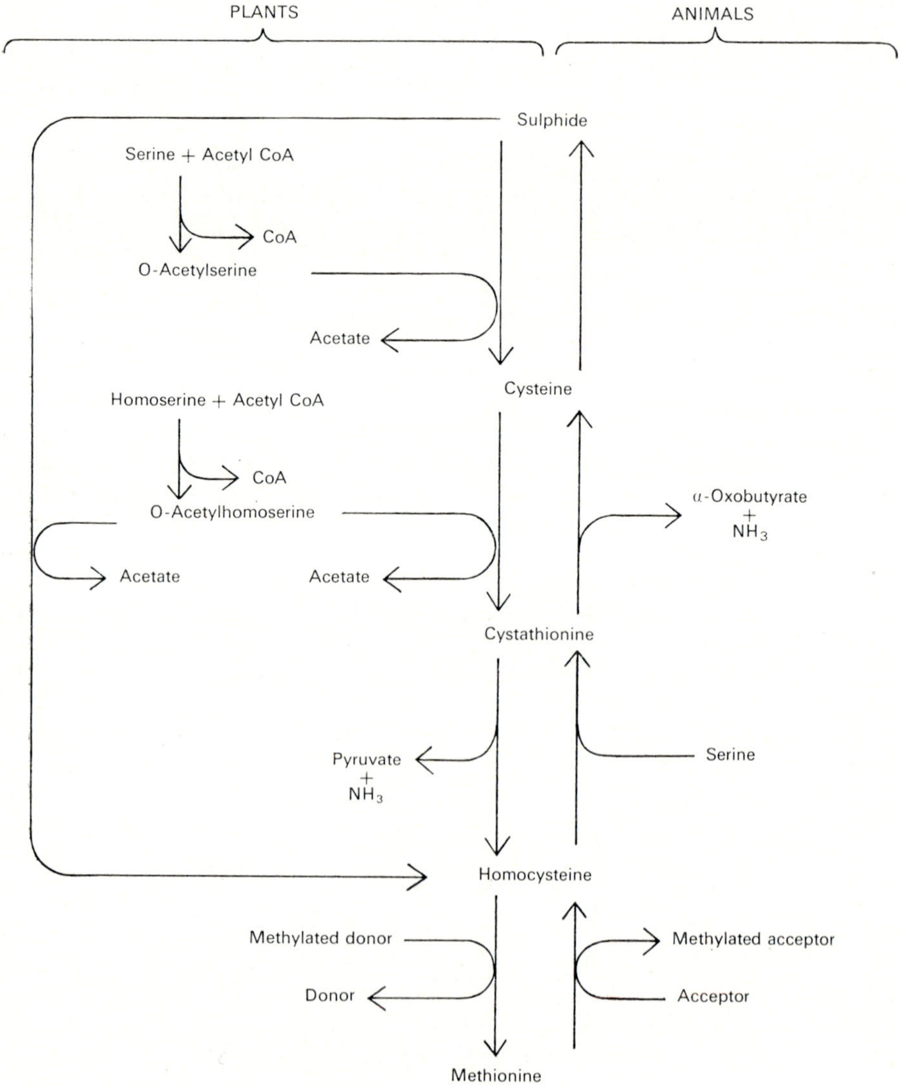

Figure 3-3 The biosynthesis of cysteine and methionine from sulphide by plants. The pathway of methionine catabolism in animals is also shown.

The synthesis of methionine from homocysteine involves the transfer of a methyl group from a methyl donor (transmethylation). In animals, derivatives of N^5-methyltetrahydrofolic acid serve as the methyl donor in a complex series of coupled reactions involving S-adenosylmethionine and the cobalt containing vitamin B_{12}. With the exception of the nodulated legumes, however, there has been no clear demonstration of a cobalt requirement for plant growth, suggesting that it is unlikely that the transmethylation mechanism of plants involves vitamin B_{12}. In some micro-organisms a second transmethylation mechanism has been described which utilizes a derivative of N^5-methyltetrahydrofolic acid as the methyl donor, but does not involve vitamin B_{12} as an intermediate. It has been suggested that an analogous reaction might be operative in plants (28).

Other reactions involving the methylation of homocysteine have been described both in plants and some micro-organisms (28); these reactions involve methionine and S-adenosylmethionine as methyl donors:

6. Homocysteine + S-adenosylmethionine $\longrightarrow$ Methionine + S-adenosylhomocysteine;
7. Homocysteine + S-methylmethionine $\longrightarrow$ 2 Methionine.

These reactions, however, cannot bring about a net synthesis of methionine since the methyl donors are themselves synthesized from methionine.

Another mechanism for the synthesis of methionine has been described in plants which does not involve transmethylation of homocysteine (11). The reaction is analogous to some of the previous reactions in that it requires O-acetyl activated homoserine as substrate; methylmercaptan serves as the donor:

8. O-acetlyhomoserine + methylmercaptan $\longrightarrow$ Methionine + acetate.

The importance of this reaction in methionine biosynthesis cannot be established until the biosynthetic origin of methylmercaptan (a common metabolite in at least some plants) is established; the reaction has no significance in the biosynthesis of methionine unless methylmercaptan is derived from sources other than methionine.

THE CELLULAR AND SUBCELLULAR LOCALIZATION OF SULPHUR METABOLISM IN PLANTS

The chloroplasts of leaves appear to be the main site of sulphate reduction in plants. The sulphate activating enzymes ATP sulphurylase and APS kinase and sulphite reductase activity are associated with chloroplasts (3,5,18). There is some evidence that reduced ferredoxin, a product of the photochemical reactions of chloroplasts, serves as the electron donor for sulphite reduction (27). The subcellular localization of the enzymes catalyzing the synthesis of O-acetylserine and cysteine in leaf tissue has not been studied. Isolated spinach chloroplasts, however, incorporate sulphur from sulphate into cysteine and the rate of incorporation was enhanced by serine (5). It is yet to be established whether cysteine biosynthesis is mediated in chloroplasts via serine sulphhydrase or O-acetylserine sulphhydrase.

Non-photosynthetic tissues also catalyze the incorporation of sulphur from sulphate into cysteine (7,21) although the amount of cysteine synthesized by the non-photosynthetic tissues of a plant is small in comparison with that synthesized by the whole plant (21). ATP sulphurylase, sulphite reductase, serine transacetylase and O-acetylserine sulphhydrase have been reported in non-photosynthetic tissues (8,18,25) where presumably reduced ferredoxin could not act as the electron donor. Serine transacetylase and sulphite reductase activities are associated with particulate fractions in non-photosynthetic tissue (18,25). These fractions have been designated mitochondrial fractions, but this designation should be viewed with caution until the particles are purified since non-photosynthetic tissue contains a wide range of organelles of similar size and density to mitochondria. The need for such caution is well illustrated by the nitrite reductases of non-photosynthetic tissues. The enzyme from wheat roots was apparently associated with a mitochondrial fraction prepared by differential centrifugation. However, when the mitochondrial fraction was further purified by density gradient centrifugation (6) the activity was shown to be associated with a particle of isopycnic density of 1.22 g cm^{-3} (proplastids?) and not with mitochondria.

The cellular and subcellular distribution of enzymes involved in the biosynthesis of methionine has not been established.

REFERENCES

(1) Asada, K., Tamura, G. and Bandurski, R. S. (1969). *J. Biol. Chem.*, **244**: 4904.
(2) Asahi, T. (1964). *Biochim. Biophys. Acta*, **82**: 58.
(3) Balharry, G. J. E. and Nicholas, D. J. D. (1970). *Biochim. Biophys. Acta*, **220**: 513.
(4) Brüggeman, J., Schlossman, K., Merckenschlager, M. and Waldschmidt, M. (1962). *Biochem. Z.*, **335**: 392.
(5) Burnell, J. N. and Anderson, J. W. (1973). *Biochem. J.*, **134**: 565.
(6) Dalling, M. J., Tolbert, N. E. and Hageman, R. H. (1972). *Biochim. Biophys. Acta*, **283**: 513.
(7) Ellis, R. J. (1963). *Phytochemistry*, **2**: 129.
(8) Ellis, R. J. (1969). *Planta*, **88**: 34.
(9) Giovanelli, J. and Mudd, S. H. (1966). *Biochem. Biophys. Res. Commun.*, **25**: 366.
(10) Giovanelli, J. and Mudd, S. H. (1967). *Biochem. Biophys. Res. Commun.*, **27**: 150.
(11) Giovanelli, J. and Mudd, S. H. (1968). *Biochem. Biophys. Res. Commun.*, **31**: 275.
(12) Giovanelli, J. and Mudd, S. H. (1971). *Biochim. Biophys. Acta*, **227**: 654.
(13) Giovanelli, J., Owens, L. D. and Mudd, S. H. (1973). *Pl. Physiol.*, **51**: 492.
(14) Hitchcock, C. and Nicholls, B. W. (1971). *Plant Lipid Biochemistry*. Academic Press, London and New York.
(15) Hodson, R. C., Schiff, J., Scarsella, A. J. and Levinthal, M. (1968). *Pl. Physiol.*, **43**: 563.
(16) Kjaer, A. (1963). In Swain, T. (ed.), *Chemical Plant Taxonomy*. Academic Press, London and New York, p. 453.
(17) Kline, B. C. and Schoenhard, D. E. (1968). *Bacteriol. Proc.*, 141.
(18) Mayer, A. M. (1967). *Pl. Physiol.*, **42**: 324.
(19) Mercer, E. I. and Thomas, G. (1969). *Phytochemistry*, **8**: 2281.
(20) O'Colla, P. S. (1962). In Lewin, R. A. (ed.), *Physiology and Biochemistry of Algae*. Academic Press, London and New York, p. 337.
(21) Pate, J. S. (1965). *Science*, **149**: 2281.
(22) Roy, A. B. and Trudinger, P. A. (1970). *The Biochemistry of Inorganic Compounds of Sulphur*. Cambridge University Press, London.

(23) Schmidt, A. (1972). *Arch. Mikrobiol.*, **84**: 77.
(24) Shaw, W. H. and Anderson, J. W. (1972). *Biochem. J.*, **127**: 237.
(25) Smith, I. K. (1972). *Pl. Physiol.*, **50**: 477.
(26) Smith, I. K. and Thompson, J. F. (1971). *Biochim. Biophys. Acta*, **227**: 288.
(27) Tamura, G., Asada, K. and Bandurski, R. S. (1967). *Pl. Physiol.*, **42**, Supp. S 36.
(28) Thompson, J. F. (1967). *A. Rev. Pl. Physiol.*, **18**: 59.
(29) Thompson, J. F. and Moore, D. P. (1968). *Biochem. Biophys. Res. Commun.*, **31**: 281.
(30) Tsang, M-L. S., Goldschmidt, E. E. and Schiff, J. A. (1971). *Pl. Physiol.*, **47**, Supp. 20.
(31) Virtanen, A. I. (1965). *Phytochemistry*, **4**: 207.
(32) Wilson, L. G. (1962). *A. Rev. Pl. Physiol.*, **13**: 201.

Sulphur Requirements of Plants*

K. SPENCER

INTRODUCTION

Sulphur is required by plants for the synthesis of certain essential amino acids and hence the elaboration of proteins, for the activation of certain enzymes, the production of certain vitamins, and for the synthesis of glucoside oils. Evidence is accumulating for other roles, most of which seem to be associated with properties of the sulphhydryl group (2,5,44).

The sulphur requirement of a crop has been defined as 'the minimum uptake of this nutrient associated with the maximum yield of dry matter' (61) and in these terms is the total amount of sulphur which enters the plant during the growing period of an annual plant, or during the growth cycle of a perennial plant. However, as growth rate changes with the development of the plant, so the requirement, when expressed as the amount of S required per unit time, can also be expected to change with plant age.

Although 'maximum' yield is specified as the level at which the requirement is measured, it is impractical in the field culture of plants to seek maximum production because of the shape of response curves. Some compromise is necessary and a proportion of the maximum, e.g., 90 per cent, is often nominated as a more realistic goal. Where economic analysis is possible, it sets the particular value. The agriculturalist is concerned with the commercially-valuable part of the plant, and the yield of this product may not be directly related to dry matter accumulation. Furthermore, since most plant products are destined as food for man or animals, quality of the product can be as important as yield. Sulphur is known to influence many aspects of plant quality (8). Thus, the ultimate objectives in meeting the sulphur needs of plants generally, may be many and varied and, as will be seen later, the precise specification of sulphur requirement requires reference *inter alia* to the species (or possibly the cultivar) and the plant part concerned, the potential yield as determined by the supply of other nutrients, and the environmental conditions.

So far requirement has been considered as the amount of sulphur necessary for the plant to absorb over some given period of time. This infers that when the requirement is just being met, sulphur compounds will occur inside the plant at a particular concentration. This concentration, the minimum amount of S per unit of dry weight associated with maximum growth by

* This review covers information published since 1965. For comprehensive accounts of significant contributions prior to 1965, the reader is referred to three earlier reviews (7,42,61).

the plant, is known as the 'critical percentage'. Provided the environmental conditions can be specified and related to experimentally-determined values, the critical percentage can be used to establish that adequate S is presently available.

Requirement, in a given situation, can also be considered in terms of how much S must be *provided* so that the plant grows optimally. The simplest situation would be that of solution culture, wherein growth is measured in a range of media concentrations for definition of that concentration supporting optimum growth. However, the problems of setting a value on the requirement of soil-grown plants and of calculating how much fertilizer S should be applied to a particular crop are complex and require some quantification of net S contribution by the soil, accessions in rainfall and losses by leaching.

THE INTERNAL AND EXTERNAL REQUIREMENTS OF PLANTS

Loneragan (38) emphasized that the term 'nutrient requirement' is used to describe two distinct relations between growth and nutrient concentration. They are the relation between growth and the *internal* concentration of the element within the plant, and the relation between growth and the *external* concentration of the element in the soil or culture solution. The two relations, and the concept of a critical value for each, are in fact the bases of plant analysis and soil testing, respectively. He also drew attention to the fact that a plant with a high internal requirement may have a low requirement for nutrient concentration in the external solution, as the capacity for uptake and distribution of nutrients can differ greatly between plants. The mobility of sulphur within plants is somewhat restricted (12), resulting in a pattern of S distribution which is characteristic of supply conditions. Sites of metabolic activity may thus suffer a shortage of S whilst older plant parts contain supraoptimal amounts.

Internal requirement

The critical percentage of one or more forms of S in a plant, provides an indirect measure of just-adequate nutrition. Published data indicate that separate values need to be established for particular growing conditions, particular stages of growth and particular plant parts. However, the expectation is that many of the problems of availability and uptake of S which render the interpretation of soil analyses difficult, are avoided.

In recent years, considerable effort has been devoted to the determination of the critical S concentration in a range of plants (14) and work on forage crops was reviewed in detail very recently (43). The sulphur fractions commonly measured are total S, sulphate-S or reducible S, protein S and non-protein organic S. In addition, total N:total S, total P:total S and protein N:protein S ratios have been used to characterize the S status of plants.

In Table 3-4, a summary of published critical percentages for a number of plants is presented. For all species for which there are several reports, there is a considerable range in values. Consideration of the individual values makes it clear that plant and environmental factors are contributing to the variation, and none of the S fractions so far examined can be considered independent of such factors. A further but not so obvious influence precluding consistency

TABLE 3-4
Reported critical S percentages for some agricultural plants (dry wt basis)

Plant	Plant part	Critical percentage or ratio*			References
		total S	SO_4—S	N:S	
Clovers	Tops (mostly)	0.15–0.30	0.017–0.040	15–20	Collated by (43), also (19), (40)
Lucerne	Tops (mostly)	0.20–0.30	0.015–0.033	10–15	Collated by (43)
Stylo	Leaves	0.12			(35)
Peanuts	Leaves	0.20–0.25			(11)
Grasses	Leaves (mostly)	0.14–0.30	0.01–0.05		Collated by (43)
Sugar cane	Leaves	0.16		17	(23)
Sweet corn	Ear leaf	0.24			(18)
Oats	Tops	0.15	0.05	20	(10)
Coconut	Nut water	0.001			(59)
	Young fronds	0.02			(59)
Coffee	Leaves		0.02		(20)

* The range in critical percentages is largely a reflection of the stage of maturity, but N status and tissue sampled are also influential.

is the great variety of analytical procedures which have been used for the separation and the determination of the S fractions. To date, much of the more detailed information has been derived under controlled environment conditions and it remains to be shown that critical percentages derived in this way are directly applicable to the fluctuating atmospheric and edaphic conditions of the farm.

External requirement

The external requirement can be described in terms of a just-adequate concentration of sulphur in the solution bathing the roots. In soil, the root's natural habitat, the concentration of available S in the rhizosphere solution varies from point to point and fluctuates with time, reflecting differences between uptake and 'fixation' on the one hand and solution, mineralization and diffusion on the other. It should be possible to obtain the average concentration to which the plant is exposed at any one time by analyzing the soil solution. For example, the solution from the larger pores can be removed from soil at field capacity by centrifugation, but there appear to be no studies relating growth to the sulphate content of solutions obtained in this way. The most widely used technique for studying the growth of plants exposed to a range of known external concentrations is solution culture.

Many of the older nutrient solutions recommended for growing plants contained up to 100 ppm S as sulphate, but by the 1950s the concentration had been reduced to about 32 ppm S (27). Recent experience suggests that even the latter level would be adequate for most plants for a considerable period without replacement of the media. When the response curve has been carefully examined at the low end, 3 to 5 ppm S has generally proved adequate for the growth of many species (see Table 3-5). Similarly, when irrigation water contained about 3 ppm S, no S deficiency was observed in field-grown sweet corn (18), nor in kikuyu grass when the soil solution contained about 5 ppm S (24). Notable exceptions to the 4 ppm ($\frac{1}{8}$ m*M*) requirement are rape and lucerne, but it is concluded that, in general, the external requirement for S in solution cultures is of the same order as for phosphorus (30).

When plants are growing in solution cultures providing the external requirement of S and in filtered air, the corresponding uptake approaches the total S requirement of that plant. Species can be compared on this basis but a requirement derived in such a way provides little guidance for those concerned in recommending fertilizer rates for field-grown crops.

THE FERTILIZER REQUIREMENT

Field experiments attempt to duplicate commercial conditions of growth by allowing the whole interplay of edaphic and atmospheric factors. By adding a series of graduated doses of nutrient an attempt is made to define the relationship between growth and added S under a particular combination of these factors thereby enabling an optimum rate of fertilizer to be nominated. The magnitude of the fertilizer requirement will obviously vary quite considerably from one experimental location to another, even though the total requirement for a given yield under specified conditions can be assumed to be constant. The fertilizer requirement is in fact a total requirement *less* the amount gained

TABLE 3-5
The external requirement for sulphur of a number of crop plants, measured in solution cultures

Crop	S concentration (ppm)	Reference	Growing conditions (litres/plant, renewal frequency, growth phase)
Rape	8	*	½ 1, weekly, veg. stage
Lucerne	8	(65)	2 1, nil, for 6 weeks
Bur clover	4	(34)	2 1, nil, to early flowering
Spanish clover	4	(31)	2 1, nil, to early flowering
Ryegrass	4	(64)	2 1, nil, for 5 weeks
Peanuts	4	(45)	4 1, to flowering
Subterranean clover, Beans, Spinach	4	**	½ 1, weekly, veg. stage
Onions	3	(21)	
Wheat	0.125	(52)	In flowing cultures, young seedlings

* P. J. Randall, priv. comm.
** D. Bouma, priv. comm.

from the air and water and the amount available in the root zone of the soil. In most situations this difference must be augmented by an appropriate allowance for non-recovery of the fertilizer, to cover that proportion of the fertilizer which is leached, fixed or metabolized by micro-organisms and weeds.

It is not surprising, therefore, that a wide range of values has been reported for each crop, a range which is largely unexplained. Precise recommendations for the use of S fertilizers are thus not possible at this stage and the range can only be narrowed if workers will endeavour to estimate the S gained or lost to the system by the various avenues.

Many of the difficulties in arriving at an accurate estimate of the S fertilizer requirement for a particular plant are concerned with putting values on the effect of the following influences:

1. the contribution to the system by sulphate in rain water and irrigation water, or by atmospheric SO_2;
2. the resultant between mineralization of S from organic matter and metabolism of S by micro-organisms (see (22));
3. the contribution of S from the various layers of the soil profile, which depends upon the rooting habit of the species;
4. the removal of sulphur from available forms by physicochemical or chemical processes, e.g., reduction during waterlogging, sorption, precipitation and leaching (see (6));

5. the interactions with other elements, which may be antagonistic, e.g., selenium, molybdenum, or synergistic, e.g., nitrogen (62);
6. the effect of environmental factors which influence uptake by the plant, e.g., temperature (9,25,26,39).

THE FERTILIZER REQUIREMENT OF PARTICULAR CROPS

For a given set of environmental conditions, the near-maximum growth of field-grown plants may be identified directly in terms of yield, or indirectly by using critical percentages of S in the plant or critical values of available S in the soil. All three bases have been utilized in compiling data on the fertilizer requirement of a number of crops (Table 3-6). There are wide ranges in the reported amounts of fertilizer S needed in deficient areas but some grouping of plants according to the magnitude of their need seems justified.

TABLE 3-6
A tentative classification of crops according to their S fertilizer requirement

Crop	Fertilizer reqt in deficient areas* kg S/ha	References
GROUP I (high)		
Cruciferous forages	40 – 80	(48), (54)
Lucerne	30 – 70	(36), (42), (51)
Rapeseed	20 – 60	(16), (41)
GROUP II (moderate)		
Coconuts	– 50	(63)
Sugar cane	20 – 40	(23), (57)
Clovers and grasses	10 – 40	(4), (19), (40)
Coffee	20 – 40	(20), (32)
Cotton	10 – 30	(3), (13), (32), (53)
GROUP III (low)		
Sugar beet	15 – 25	(54)
Cereal forages	10 – 20	(10)
Cereal grains	5 – 20	(7), (46)
Peanuts	5 – 10	(11), (28)

* Figures cited for the high end of the range apply where the potential yield is high, accessions in rainfall are low, the soil is low in available S and there is considerable loss in effectiveness of applied S. Figures cited for the low end refer to the opposite situation.

For perennials, a further consideration is whether the requirement refers to a corrective fertilizer dressing or to an annual maintenance dressing. The former is typically about four times the latter in amount.

Some recent observations concerned with the S requirement of particular crops are summarized below. Most comments deal with influences which have been shown to cause variations in the fertilizer requirement.

Grain crops

Cereals

S responses by cereals occur in restricted areas throughout the world (7). None have been recorded in Australia, and the inference has been that requirement is low. However, recent work (29,46) has suggested that the absolute requirement of cereals may be similar to that of legumes but that a greater efficiency in uptake (or sometimes a restriction by low moisture or N) militates against field responses.

Forage crops

Grasses

Solution culture work, e.g., (64) suggests that grasses have a low to moderate S requirement but little information is available for pure grass stands. In general, mixed pastures are fertilized according to the needs of the legume component so that N-fixation is maximized. In such cases, the S needs of the grasses appear to be met (66,67).

Lucerne

The available evidence suggests a moderate to high fertilizer requirement for lucerne.

The requirement of lucerne for S has been shown to be affected by productivity, leaching, time of year, and cultivar. Californian and Oregon experience have been contrasted (42) in respect of productivity of the stand; only about half the Californian requirement was needed in Oregon. In France, limestone soils (51) are considered to need more S than other soils, probably because negligible adsorption results in greater leaching at high pHs. Requirement may vary throughout a season, the variation in a southern Queensland trial being attributed to a seasonal wave in mineralization rather than to leaching (36). Of five lucerne cultivars, 'du Puits' grew the poorest without S addition and responded the most to added S (15).

Clovers

External requirements (Table 3-5), critical percentages in tissues (Table 3-4) and fertilizer experiments (Table 3-6) indicate a moderate requirement for S by clovers.

As with lucerne, seasonal conditions influence productivity and therefore requirement of clovers (4). The particular cultivar (33) also influences requirement.

Of special significance with certain clovers is the content of oestrogenic flavones. Opposite trends in content with S supply have been shown in subterranean clover and red clover (55,56).

Others

Saalbach (54) has calculated the amount of S required by several European forage crops at a range of levels of yield expectation. The relatively high demand by rape was clearly shown and solution cultures (Table 3-5) confirm this assessment.

Grazed cereals may be expected to have a somewhat higher requirement than those grown for grain. Not only are forage cereals grown in wetter regions, with poorer soils and greater losses of added S, but defoliation precludes the accumulation of reserves by the plant. An appreciable requirement by grazed oats fertilized with nitrogen was demonstrated on the Central Tablelands of N.S.W. (10,17).

The low S content of cereal forages is often of concern and the difficulty of increasing the content was demonstrated with five varieties of sorghum (49). They were grown on a soil in which S was adequate (for yield of dry matter) and fertilized with 45 kg S per ha, but not one reached an S content of 0.15 per cent, a level considered optimum for beef cattle.

Sugar crops

Sugar cane

In 1966, Stanford and Jordan (61) were unable to find factual information on this crop but a severe deficiency has since been reported from Rhodesia (23). A moderate requirement for S was indicated, both weight of cane and sucrose content being increased.

Only one slight response in eight extended trials was reported from Queensland (57). Rainfall samples showed that accessions over the period were not sufficient to account for the lack of response.

The important considerations in anticipating a need for fertilizer S in Hawaii are all concerned with sulphate retention by the soil. Deficiency is stated to be more pronounced if pH values are > 6, the soils are sandy, strongly leached and with shallow profiles (70).

Sugar beet

Whitehead's assertion that sugar beet has a relatively high requirement (68) is not borne out by a subsequent report (51).

Fibre crops

Cotton

In summarizing many trials in southeastern U.S.A., Jordan (37) concluded that the soils were becoming marginally deficient for cotton culture and that the requirement as fertilizer was inversely related to the amount of S brought down in rain. A total requirement (S in rain plus S in fertilizer) of about 15 kg/ha was indicated for that area. In a more S-deficient region, the campos cerrados soils of Brazil, about twice that amount is needed (32). As in the western cotton areas of U.S.A., no published reports of S responses have come from the Ord River or Narrabri areas in Australia. The requirement seems to be met from soil reserves, irrigation waters and the customary basal application of single superphosphate.

In equatorial Africa, the S requirement was found to decrease with time since clearing the bush (13). A similar trend was found with peanuts (50) and was explained in terms of greater mineralization as the C:S ratio in the soil narrowed.

Oil crops

As oil storage organs are usually also rich in protein, it has been assumed that the S requirements of oil seed crops would be high. This appears to be true for rapeseed but not for peanuts or soybeans.

Peanuts

A modest requirement in fertilizer was found in central Africa, response occurring in yield of hay and pods, and in protein content but not oil content (11). A larger requirement is indicated in New Guinea (28).

Soybeans

Remarkably little work on S nutrition of this important crop has been published (1). A greenhouse study (58) showed yield increases in plants as solution concentration of S increased up to 10 ppm whereas protein content (mainly as methionine and lysine) increased up to 20 ppm.

Nitrogen-fed plants required twice the S concentration in solution cultures to reach maximum growth as did nodulated plants (69).

Rapeseed

The high requirement of this plant has already been mentioned, responses up to 60 kg S/ha having been recorded (16,47,48) when nitrogen status was adequate. The high requirement can aggravate practical difficulties in maintaining an adequate supply, e.g., in France, split dressings in autumn and late winter, are necessary (41).

Coconuts

In the last ten years, S deficiency of coconuts has been recognized as an important restriction on production of copra in Papua-New Guinea. Not only the number of nuts is reduced but the oil content and quality thereof is detrimentally affected (59). It has subsequently been shown that sulphur affects oil synthesis in the maturing nut (60). About ½ kg S/palm has been recommended (63).

REFERENCES

(1) Adams, C. A. and Johnson, R. E. (1968). *Pl. Physiol.*, **43**: 2041.
(2) Adepipe, N. O., Hofstra, G. and Ormrod, D. P. (1972). *Can. J. Bot.*, **50**: 1789.
(3) Anderson, O. E. and Futral, J. G. (1966). Georgia Agric. Exp. Sta. Bull. N.S., p. 167.
(4) Atkinson, W. T. (1966). *Proc. IX Int. Grassld Congr.*, p. 655.
(5) Banath, C. L. and Single, W. V. (1971). *Agric. Gaz. N.S.W.*, **82**: 307.
(6) Barrow, N. J. (1974). *Sulphur in Australasian Agriculture*, McLachlan, K. D. (ed.), Sydney University Press, Sydney, p. 50.
(7) Beaton, J. D. (1966). *Soil Sci.*, **101**: 267.
(8) Beaton, J. D. and Fox, R. L. (1971). 'Production, marketing, and use of sulphur products', in Olson, R. A., Army, T. J., Hanway, J. J. and Kilmer, V. J. (eds), *Fertilizer Technology and Use*. 2nd edn, Soil Sci. Soc. Amer., Madison, Wis.
(9) Blair, G. J. (1970). *Comm. Soil Sci. Pl. Anal.*, **1**: 115.
(10) Blair, G. J. and Crofts, F. C. (1970). *Aust. J. Exp. Agric. Anim. Husb.*, **10**: 768.
(11) Bockelee-Morvan, A. and Martin, G. (1966). *Sulphur Inst. J.*, **2**(2): 2.
(12) Bouma, D. (1967). *Aust. J. Biol. Sci.*, **20**: 613.
(13) Braud, M. (1969-70). *Sulphur Inst. J.*, **5**(4): 3.

(14) Chapman, H. D. (1966). *Diagnostic Criteria for Plants and Soils.* University of California, Div. of Agric. Sciences.
(15) Chapman, F. M., Mason, J. L. and Miltmore, J. E. (1972). *Can. J. Pl. Sci.*, **52**: 493.
(16) Courpron, C., Menet, M. and Tauzin, J. (1972). *Sulphur Inst. J.*, **8**(1 & 2): 15.
(17) Crofts, F. C. and Blair, G. J. (1966). *J. Aust. Inst. Agric. Sci.*, **32**: 142.
(18) Daigger, L. A. and Fox, R. L. (1971). *Agron. J.*, **63**: 729.
(19) Dawson, M. D. (1969). *Sulphur Inst. J.*, **5**(3): 16.
(20) De Freitas, L. M. M., Pimental Gomes, F. and Lott, W. L. (1972). *Sulphur Inst. J.*, **8**(3): 9.
(21) Freeman, G. G. and Mossadeghi, N. (1970). *Proc. Biochem. Soc.*, **118**: 24.
(22) Freney, J. R. (1974). *Sulphur in Australasian Agriculture.* McLachlan, K. D. (ed.), Sydney University Press, Sydney, p. 31.
(23) Gosnell, J. M. and Long, A. C. (1969). *Proc. S. Afr. Sugar Technol. Assoc.*, **43**: 26.
(24) Hasan, S. M., Fox, R. L. and Boyd, C. C. (1970). *Proc. Soil Sci. Soc. Am.*, **34**: 897.
(25) Herath, H. M. V. and Ormrod, D. P. (1971). *Pl. Soil*, **35**: 635.
(26) Herath, H. M. V. and Ormrod, D. P. (1973). *Pl. Soil*, **38**: 525.
(27) Hewitt, E. J. (1966). Comm. Agric. Bur. Tech. Comm., p. 22.
(28) Hill, G. D. (1970). *Papua N.G. Agric. J.*, **22**(1): 26.
(29) Holford, I. C. R. (1971). *Aust. J. Agric. Res.*, **22**: 879.
(30) Hossner, L. R., Freeouf, J. A. and Folsom, B. L. (1973). *Proc. Soil Sci. Soc. Am.*, **37**: 405.
(31) Hylton, L. O., Cornelius, D. R. and Ulrich, A. (1968). *J. Range Mange.*, **21**: 129.
(32) Jones, M. B. (1967). *Sulphur Inst. J.*, **3**(2): 2.
(33) Jones, M. B., Lawler, P. W. and Ruckman, J. E. (1970). *Agron. J.*, **62**: 439.
(34) Jones, M. B., Ruckman, J. E. and Lawler, P. W. (1972). *Agron. J.*, **64**: 55.
(35) Jones, R. K., Robinson, P. J., Haydock, K. P. and Megarrity, R. G. (1971). *Aust. J. Agric. Res.*, **22**: 885.
(36) Jones, R. M. (1970). *Aust. J. Exp. Agric. Anim. Husb.*, **10**: 749.
(37) Jordan, H. V. (1964). *USDA Tech. Bull.*, 1196.
(38) Loneragan, J. F. (1968). *Nature* (London), **220**: 1307.
(39) McKell, C. M. and Wilson, A. M. (1963). *Agron. J.*, **55**: 134.
(40) McLachlan, K. D. and De Marco, D. (1973). *Aust. J. Exp. Agric. Anim. Husb.*, **13**: 75.
(41) Maraby, T. (1968). *Sulphur Inst. J.*, **4**(2): 2.
(42) Martin, W. E. and Walker, T. W. (1966). *Soil Sci.*, **101**: 248.
(43) Metson, A. J. (1973). *Sulphur Inst. Tech. Bull.*, 20.
(44) Netolitzky, H. J. and Skoropad, W. P. (1971). *Can. J. Pl. Sci.*, **51**: 49.
(45) Nicholaides, J. J. and Cox, F. R. (1970). *Agron. J.*, **62**: 262.
(46) Nyborg, M. (1968). *Can. J. Soil Sci.*, **48**: 37.
(47) Nyborg, M. and Bentley, C. F. (1971). *Sulphur Inst. J.*, **7**(3): 16.
(48) Odelien, M. (1966-7). *Sulphur Inst. J.*, **2**(4): 14.
(49) Owen, D. F. and Furr, R. D. (1967). *Agron. J.*, **59**: 611.
(50) Prevot, P. and Ollagnier, M. (1964). *Agrochimica*, **8**: 211.
(51) Radet, E. (1966). *Sulphur Inst. J.*, **2**(1):11.
(52) Reisenauer, H. M. (1968). *Agron. Abs.*, p. 108.
(53) Rogers, H. V. (1969). *Cotton Farming*, **13**(2): 46.
(54) Saalbach, E. (1965-6). *Sulphur Inst. J.*, **1**(2): 7.
(55) Schoo, H. and Rains, D. W. (1971). *Crop Sci.*, **11**: 716.
(56) Schultz, G. (1969). *Soils Fertil.*, **32**(4): 400.
(57) Sedl, J. M. (1971). Bur. Sugar Exp. Sta., Q. Tech. Comm. 2.
(58) Sharma, G. C. and Bradford, R. R. (1973). *Comm. Soil Sci. Pl. Anal.*, **4**: 77.
(59) Southern, P. J. (1967). *Papua N.G. Agric. J.*, **19**: 38.
(60) Southern, P. J. (1969). *Oleagineux*, **24**: 211.
(61) Stanford, G. and Jordan, H. V. (1966). *Soil Sci.*, **101**: 258.
(62) Stewart, B. A. and Porter, K. L. (1969). *Agron. J.*, **61**: 267.

(63) Sumbak, J. H. (1971). *Papua N.G. Agric. J.*, **22**: 77.
(64) Ulrich, A. and Hylton, L. O. Jr (1968). *Pl. Soil*, **29**: 274.
(65) Ulrich, A., Tabatabai, M. A., Ohki, K. and Johnson, C. M. (1967). *Pl. Soil*, **26**: 235.
(66) Walker, T. W. and Adams, A. F. R. (1958). *Pl. Soil*, **9**: 353.
(67) Walker, T. W., Adams, A. F. R. and Orchiston, H. D. (1956). *Pl. Soil*, **7**: 290.
(68) Whitehead, D. C. (1964). *Soils Fertil.*, **23**: 1.
(69) Wooding, F. J., Paulsen, G. M. and Murphy, L. S. (1970). *Agron. J.*, **62**: 277.
(70) Yonemitsu, I. (1967). Rep. Hawaiian Sugar Technol., **26**: 24. (in Trop. Abs., **23**: 519, 1968.)

The Role and Requirement for Sulphur in Ruminant Nutrition

R. J. MOIR

The animal industries of Australia are based largely on grazing ruminants. The feed industries are relatively small and confined mainly to poultry, pig and dairy cow feed preparations. The soil-plant-animal relationships are therefore basic to not only understanding the bases of production from ruminants, but vital to production itself. The major products of the pastoral industries, meat and wool, are both protein-rich materials and therefore dependent upon the nitrogen and sulphur train from air and soil to plant to animal to product. The relatively ancient landscape of this country is notoriously deficient in the critical ingredients of the principal animal products. The extensive systems using natural pastures depended upon relatively slow rates of nitrogen accretion and soil sulphur mineralization. These are giving way to improved pastures with high nitrogen fixing potential under legumes made possible by soil amelioration with the phosphates and, fortuitously perhaps, the sulphates of superphosphate, together with required trace elements.

The drastic seasonal conditions prevailing over most of the continent, resulting in rapid maturation of forage species which then become the major or sole source of nutrients for grazing animals, present great difficulties to management and continuous production.

The protein nutrition of conventional animal organisms is comparatively simple; given adequate, but easily accessible, energy, vitamins and minerals, most proteins from actively metabolizing tissues, animal or vegetable, suffice. With few exceptions their nutritive value can be improved by small additions of methionine, the primary sulphur amino acid of animal tissue, and the universal first limiting amino acid in animal nutrition. Plant storage proteins, however, present problems as, in addition to methionine, one or more of the other eight or ten essential amino acids are frequently limiting, notably lysine. Methionine alone can virtually satisfy the total sulphur requirements of animal tissues, although the requirement can be reduced by cystine and to some extent by sulphate for sulphation processes.

Ruminant animals, because of the interpolation of the micro-organisms of the rumen between the food sources and the animal tissues, present a much more complicated protein nutritional system. The amino acid constitution of the diet and of the quantities of amino acids present in the fodder does not necessarily indicate the nutritive value. The concept of a biological value of a diet has, in fact, a greater validity than that of the biological value of a protein.

The anaerobic, well buffered, continuous flow fermentative system that constitutes the rumen, maintains a dense and varied microbial population. Proteolytic and deaminative activity of this microflora removes a variable, but usually, substantial proportion of the dietary protein from the food chain, so that the protein eventually available to the animal depends upon the undigested proportion of dietary protein flowing from the rumen in the effluent from that organ, together with microbial and protozoal protein synthesized during the fermentation process. The level of amino acids in the rumen is never high (35) and is not considered to be a significant factor in the nutrition of the animal: the low level is undoubtedly due to the rapid catabolism of amino acids in the rumen.

THE DEGRADATION OF SULPHUR AMINO ACIDS

That protein could be extensively degraded in the rumen was clearly demonstrated by the work of McDonald (37,38) showing the production of ammonia in the rumen. Other work (21), using *in vitro* washed suspensions of ruminal micro-organisms, showed the production of fatty acids, including branch chain fatty acids, from protein degradation. Amino acids are not degraded at the same rate; the order of degradation of the 22 amino acids tested varied with dietary conditions (49). There were clear indications from this work that cystine and cysteine were degraded very rapidly compared with methionine. This pattern of degradation, based on ammonia production *in vitro* (35,36,45), has been confirmed with other *in vitro* studies on sulphide production (11,30).

Incubating rumen contents with cysteine-U-^{14}C and ^{35}S-cystine, or with with methionine-1,2,3,4-^{14}C and ^{35}S-methionine, led to the recovery of approximately 70 per cent ^{14}C as VFA and 85 per cent ^{35}S as H_2S, in the case of cysteine, but only approximately 50 per cent of the isotopes as VFA or H_2S respectively, in the case of methionine after 105 minutes (41).

In vivo, the rapid and substantial degradation of cyst(e)ine is unequivocal and is in keeping with the *in vitro* studies. The addition of 0.66 per cent cystine to a low sulphur diet (0.075 per cent) fed at two-hourly intervals raised the ruminal sulphide level from 0.54 μg/ml to 4.26 μg/ml (32). Even so a small amount of cystine passed out of the rumen without degradation. Infusions of 3 g of cystine, cysteine or methionine sulphur as a single dose into the rumen of sheep, immediately after feeding a restricted (200 g) diet, gave peak ruminal sulphide levels of approximately 45 μg/ml for both cysteine and cystine; although sulphide rose significantly with methionine, the level remained below 5 μg/ml (2). In addition, however, thiomethane was also released. The relationship of H_2S concentration (x) to CH_3SH concentration (y) was $y = 1.08 + 0.738\,x$ ($r = 0.58$ $P < 0.01$). Bird (2) suggests that the thiomethane may be demethylated to H_2S. Adaptation for twenty days with methionine infusions did not alter the situation. On the basis of a decline in CH_3SH concentration in *in vitro* experiments (55), it has been claimed that the thiomethane is utilized by ruminal bacteria; the reaction has been shown to occur in bacteria (23). Even assuming that the production of methyl mercaptan and its oxidation products may have been overlooked, or not measured in earlier work, there appear to be very large variations in methionine degradation rate. It may be inferred from several feeding studies that orally fed methionine is almost completely degraded, as either most of the amino acid sulphur, whether fed as methionine alone or as protein (19), has appeared shortly in the urine or has had little, if any, effect upon wool growth (19,45); the data of other experiments require a slightly different interpretation (29). In yet another experiment (8) where 2 g methionine was given daily per rumen or per abomasum, together with ^{35}S-methionine, expected responses were obtained; 22.8 per cent of the ^{35}S appeared in wool with abomasal infusion, but only 5.6 per cent with ruminal infusion. As in a previous study (2) sulphide and methyl mercaptan levels in ruminal and omasal contents were very low, indicating a slow rate of degradation. The digestibility and utilization of sulphur amino acids of rumen bacteria, as indicated by ^{35}S incorporation studies, is as high as those of casein (5).

Examination of the flow data of digesta through the omasum indicated that 74 per cent of the ^{35}S infused into the rumen left by this route; only 6.4 per cent was inorganic S, the remainder was organic and associated with solids. However, 39 per cent of the ^{35}S was recovered in the urine and a high 30 per cent in the faeces, indicating that most of the organic S was not unmodified methionine but possibly some deaminated, but not demethiolated product. It has been suggested that some methionine may escape the rumen because of the slower rate of degradation, particularly where the outflow of digesta is rapid (2). This has been confirmed on wool growth data in recent work (20).

Methionine hydroxy analogue (MHA) is less soluble and less readily degraded in rumen fluid *in vitro* than is methionine (47). While there is doubt that MHA can be utilized by rumen bacteria, it can be aminated by animal tissues to methionine. Orally fed MHA has given production responses in cattle (9) and in sheep, although it is more effective when infused abomasally (34).

INORGANIC SULPHATE

Considerable sulphate is frequently ingested. Sulphate may be reduced by assimilatory pathways by bacteria, or by dissimilatory mechanisms yielding free H_2S which presumably competes with methanogenesis for available hydrogen within the rumen. Dissimilatory sulphate-reducing bacteria have been isolated from the rumen (17). The dissimilative APS and assimilative PAPS systems are present in rumen bacteria (16).

Whereas desulphuration of cystine is constitutive, sulphate reduction is adaptive, and may be variable. One of four sheep failed to show enhanced sulphide levels with sulphate, but not with cystine in Hume and Bird's experiment (32). In most instances (39) adaptation is a normal state, due to sulphate in the fodder. When poorly adapted the process is usually complete within fourteen days, and levels of ruminal sulphide increase in proportion to sulphate intake (7). In preliminary experiments, with sulphate infused through the duodenum, sulphide was noted in the faeces after several days' infusion.

SULPHUR FOR RUMEN MICRO-ORGANISMS

Apart from methionine, which is degraded at variable rates, sulphur amino acids and inorganic sulphur compounds, including elemental sulphur, rapidly yield sulphide. Nader and Walker (41) have indicated that the maximum free sulphur amino acid incorporation was 1 per cent for cysteine and 11 per cent for methionine. From this and other work, it may be inferred that the central source of sulphur for ruminal microbial synthesis is sulphide. Methionine is essential or stimulatory to certain strains of the ruminal bacteria *Bacteroides ruminocola* (42) and certain *Lactobacillus* spp. (27). However, as the transfer of sulphur from methionine to cystine, as in mammals, does not usually occur in bacteria, methionine *per se* can only satisfy methionine requirements and then only if it can be absorbed by the cell; the remaining sulphur requirements must, therefore, be derived from sulphide.

The fact that elemental sulphur and sulphate can be utilized has long been established (10,50); there were, however, reports that sulphate-S was not used by many species of rumen bacteria (22). It may be expected that micro-organisms will preferentially use more reduced sulphur compounds in the medium, and the common practice of using sulphide and cysteine to achieve necessary Eh conditions in culturing ruminal bacteria may have conditioned some of the results. The probable pattern is well illustrated by the early classical work of Roberts *et al.* (46) with *E. coli*, both with respect to the preferred use of more reduced sulphur, and competition between sulphur species. *In vitro* studies (24) with rumen bacteria showed only 10 per cent of incorporated sulphur from sulphate in the presence of sulphide and approximately 20 per cent from cystine and methionine in the presence of sulphate.

SOURCES OF SULPHUR IN THE RUMEN

Dietary sulphur, whether as amino acid or inorganic sulphur, supplies the major portion of the ruminal micro-organisms requirements even under deficiency conditions. Recycling of sulphur is minimal in sheep, but may be more extensive in cattle. As little as 200 mg per day may be recycled to the rumen of sheep, mainly, if not entirely, through the saliva (3,12,13). Significantly higher levels have been reported (460 mg) (32) and the amount may be a function of blood sulphate level (15); a high correlation exists between blood sulphate and salivary sulphur ($r = 0.817$ $P < 0.001$) in cattle where minimal salivary sulphur in the absence of eating or ruminating was estimated to be 1.0 to 2.3 g per day (28). Furthermore, the ratio of nitrogen to sulphur in sheep's saliva is much wider (11-12:1) (15) than that in cattle (4:1) (28). With low sulphur content diets, 0.023 per cent S (3) and 0.075 per cent (32) there was a net increase in the amount of sulphur in the digesta leaving the rumen above intake.

SULPHUR AND RUMINAL FUNCTION

Linear growth responses to sulphur up to 10 μg S/ml have been found with pure cultures and with mixed suspensions of ruminal micro-organisms using sulphur-free basal media reduced with ascorbic acid and boron hydride (24). Sheep fed purified diets very low in sulphur fail to grow and deaths occurred (25,51,54); the animals maintained health and vigour with added sulphate.

Many feeding experiments have demonstrated effects of sulphur additions either as sulphate, elemental sulphur, methionine or cystine upon digestibility and intake; some experiments have failed to show responses in one or more of the variables measured. Some of the responses are outlined here. Where the diet is frankly sulphur deficient so that potential fermentibility cannot be expressed, added sulphur should result in firstly, increased microbial growth and hence protein synthesis (32), secondly, increased rate of digestion (15) and dry matter digestibility (3,15) and ultimately increased intake as a result of each of these factors.

Sulphate and/or cystine supplements to an 800 g oathull (50 per cent), starch (25 per cent), sucrose (12 per cent) ration containing adequate nitrogen (2.57 per cent) but marginal sulphur (0.075 per cent) increased digestibility slightly but significantly (74 to 78 per cent) and increased protein synthesis (82 g/day to 91-98 g/day) thus significantly improving nitrogen and sulphur balances (32). The rate of ruminal digestion increases in a higher sulphur environment (15), the time for 50 per cent loss of cellulose from captive bundles of cotton fell from 158 hours to 62 hours when dietary sulphur was increased from 0.058 to 0.143 per cent. A most striking effect on ruminal digestion was obtained by Bird (3), who found a 22 per cent increase in overall digestibility with a relatively small (0.36 g) addition of sulphur to a very sulphur deficient (0.025 per cent S) roughage diet. On the basal diet only 10 per cent of the organic matter was digested in the rumen compared with 75 per cent in the supplemented diet.

In most circumstances, the nitrogen and sulphur content of poor fodders parallel one another so that supplementation with nitrogen alone is ineffective. As molasses may contain appreciable sulphur, e.g. up to 1 per cent (48), its use as a vehicle for urea may have as much effect through its sulphur content as through the readily digested carbohydrates. The addition of urea to spear grass (*Heteropogon contortus*) had no influence on digestibility nor intake in either sheep or cattle (48). The addition of sulphate to the urea-containing diets substantially increased intake without significantly increasing digestibility. Sulphur with or without urea increased intake of spear grass by sheep (33). No increase in response was found above the addition of 0.12 g S/day, but some depression was observed with 0.51 g S/day as sulphuric acid. Plasma urea levels fell linearly with increasing sulphur intake, whereas plasma sulphate rose. In the earlier study, it was interesting that the plasma urea levels of cattle were much lower on the spear grass alone (5.6 mg/100 ml) in cattle than in sheep (14.5 mg/100 ml) (48). It is tempting to speculate that the possible higher return of sulphur to the rumen in cattle (28) than in sheep (12) may be responsible in part. With nitrogen adequate diets such as the legume Townsville stylo (*Stylosanthes humilis*), sulphur addition increased intake by 10 per cent (33). A rather greater effect was noted on a mixed ration of spear grass with 20 per cent Townsville stylo (43).

Because of the variations in response in the experiments above, it is clear that, in considering the effects of sulphur addition, animal factors as well as the nature, chemical composition and potential fermentibility of the foodstuff must be considered. Thus differences between sheep and cattle have been noted, as well as breed differences (48).

While sulphate *per se* is not readily absorbed from the rumen, sulphide is lost rapidly by absorption (14); sulphide levels reach high levels following infusion or ingestion of sulphate and substantial proportions may be lost (7). The total daily flow of sulphate was estimated to be only 126 mg/day, irrespective of intake, which included treatments of up to 6 g sulphate sulphur per day. In these experiments and many others, elevated sulphide following adaptation resulted in reduced intake and feed refusals. The effects of sulphide on sheep have been studied further using sulphide infusions (6). Sheep given repeated doses of sulphide show respiratory distress, muscular spasms, and collapse may occur; and temporary ruminal stasis and reduced rumination. Eructated H_2S is rapidly absorbed through the lungs and results in coma (18). The death of cows given elemental sulphur in addition to the basal sulphate of the synthetic diet used over periods of up to 394 days has been attributed to the elemental sulphur (53).

The effect of sulphide from sulphate, sulphite and amino acids may well place a limit on sulphate supplementation. It has been suggested that 4 g sulphate sulphur or 0.2 per cent SO_4-S of total diet dry matter is more than is required for adequate protein synthesis, but is less than the level likely to reduce feed intake.

A single figure for sulphur requirement cannot be readily given at this time. For roughage rations, contents below 0.1 per cent total S would appear to be marginally deficient, but with purified and concentrate diets, higher levels would appear to be desirable (26). In most circumstances, a safe approach to sulphur is to base sulphur requirement on nitrogen using an N:S ratio of 10:1 (40).

Apart from the strictly amino acid functions of sulphur, it is implicated in a wide range of other reactions. The ruminant is noted for its lack of dependence upon dietary water soluble vitamins; these are synthesized by the ruminal bacteria and the synthesis of biotin and riboflavin are stimulated by adequate sulphur levels (see (26)). Sulphate is also used in chondroitin sulphate and in the mucins of the saliva, gastro-intestinal tract, reproductive tract and other duct systems. The quantitative requirements are not known, but undoubtedly inorganic sulphate can reduce the catabolism of sulphur amino acids for sulphation purposes. The sulphur incorporated in the bile acids is effectively conserved by an efficient entero-hepatic cycle (4).

Sulphur also interacts with other nutrients, notably selenium and copper and molybdenum (see (52)). Deficiency of selenium results in white muscle disease in ruminants, and sulphate addition (0.33 per cent of diet) increases the incidence of the condition (31); it is alleviated by dosing with selenium. At low levels of sulphur (0.05 per cent) the excretion of selenium, both in faeces and urine, was significantly lower than at higher sulphur levels (0.11 to 0.24 per cent S) (44). Several factors may be involved in the Se-S interaction, such as the reduction of selenates and selenites to Se^0 and competition between the two for absorption and metabolism.

The copper-molybdenum-sulphate interaction is a complex one and operates at both the ruminal and tissue levels and has been fully reviewed (52). It has been shown recently that molybdenum influences sulphate reduction to sulphide in the rumen (16); sulphide is stimulated up to 40 ppm molybdenum and inhibited at higher levels. The level of sulphide in the rumen may have a substantial influence upon the availability of copper due to precipitation (1). Other heavy metals may be similarly affected. Sulphur may act both to the advantage or to the detriment of the animal so that requirement is very much conditioned by metabolic needs and the total chemical environment of the ruminant system.

REFERENCES

(1) Bird, P. R. (1970). *Proc. Aust. Soc. Anim. Prod.*, **8**: 212.
(2) Bird, P. R. (1972). *Aust. J. Biol. Sci.*, **25**: 185.
(3) Bird, P. R. (1972). *Aust. J. Biol. Sci.*, **25**: 1073.
(4) Bird, P. R. (1972). *Aust. J. Biol. Sci.*, **25**: 817.
(5) Bird, P. R. (1972). *Aust. J. Biol. Sci.*, **25**: 195.
(6) Bird, P. R. (1972). *Aust. J. Biol. Sci.*, **25**: 1087.
(7) Bird, P. R. and Moir R. J. (1971). *Aust. J. Biol. Sci.*, **24**: 1319.
(8) Bird, P. R. and Moir, R. J. (1972). *Aust. J. Biol. Sci.*, **25**: 835.
(9) Bishop, R. B. (1971). *Sulphur Inst. J.*, **7**(2): 9.
(10) Block, R. J., Stekol, J. A. and Loosli, J. K. (1951). *Arch. Biochim. Biophys.*, **33**: 353.
(11) Bosman, M. S. (1965). Jaarb. Inst. Biol. scheik. Onderz. LandbGewars. p. 97.
(12) Bray, A. C. (1964). *Proc. Aust. Soc. Anim. Prod.*, **5**: 336.
(13) Bray, A. C. (1969). *Aust. J. Agric. Res.*, **20**: 725.
(14) Bray, A. C. (1969). *Aust. J. Agric. Res.*, **20**: 739.
(15) Bray, A. C. and Hemsley, J. A. (1969). *Aust. J. Agric. Res.*, **20**: 759.
(16) Bryden, J. McG. and Bray, A. C. (1972). *Proc. Aust. Soc. Anim. Prod.*, **9**: 335.
(17) Coleman, G. S. (1960). *J. Gen. Microbiol.*, **22**: 423.
(18) Dougherty, R. W., Mullenax, C. H. and Allison, M. J. (1965). In Dougherty, R. W. (ed.), *Physiology of Digestion in the Ruminant*, Butterworths, Washington D.C.
(19) Downes, A. M., Reis, P. J., Sharry, L. F. and Tunks, D. A. (1970). *Brit. J. Nutr.*, **24**: 1083.
(20) Doyle, P. T. (1972). 'Methionine Utilization by Sheep'. Unpublished honours thesis.
(21) El-Shazly, K. (1952). *Biochem. J.*, **51**: 640
(22) Emery, R. S., Smith, C. K. and Fai-to, L. (1957). *Appl. Microbiol.*, **5**: 363.
(23) Flavin, M. and Slaughter, C. (1967). *Biochim. Biophys. Acta*, **132**: 400.
(24) Fountain, R. D. (unpublished data).
(25) Gall, L. S., Thomas, W. E., Loosli, J. K. and Huhtanen, C. N. (1951). *J. Nutr.*, **44**: 113.
(26) Garrigus, U. S. (1970). Chapter 9 in Muth, O. H. (ed.), *Symposium. Sulphur in Nutrition.* Avi Publishing Co., Westport, Connecticut.
(27) Gibbons, R. J. and Doetsch, R. N. (1959). *J. Bact.*, **77**: 417.
(28) Goss, K. F., Thornton, R. F. and Moir, R. J. (1969) (unpublished results).
(29) Graceva, L. V. (1969). Ovcevodstvo No. 2, p. 34. (Cited in *Nutr. Abstr.*, **39**, No. 8158).
(30) Halverson, A. W., Williams, G. D. and Paulson, G. D. (1968). *J. Nutr.*, **95**: 363.
(31) Hintz, H. F. and Hogue, D. E. (1964). *J. Nutr.*, **82**: 495.
(32) Hume, I. D. and Bird, P. R. (1970). *Aust. J. Agric. Res.*, **21**: 305.
(33) Kennedy, P. M. and Siebert, B. D. (1972). *Aust. J. Agric. Res.*, **23**: 45.
(34) Langlands, J. P. (1972). *Proc. Aust. Soc. Anim. Prod.*, **9**: 325.
(35) Lewis, D. (1955). *Brit. J. Nutr.*, **9**: 215.
(36) Lewis, D. and Emery, R. S. (1962). *J. Dairy Sci.*, **45**: 765.
(37) McDonald, I. W. (1948). *Biochem. J.*, **42**: 584.
(38) McDonald, I. W. (1952). *Biochem. J.*, **51**: 86.
(39) Moir, R. J. (1970). Chapter 11 in Muth, O. H. (ed.), *Symposium: Sulphur in Nutrition.* Avi Publishing Co., Westport, Connecticut.
(40) Moir, R. J., Somers, M. and Bray, A. C. (1967-8). *Sulphur Inst. J.*, **3**: 15.
(41) Nader, C. J. and Walker, D. J. (1970). *Appl. Microbiol.*, **20**: 677.
(42) Pittman, K. A. and Bryant, M. P. (1964). *J. Bact.*, **88**: 401.

(43) Playne, M. J. (1969). *Aust. J. Exp. Agric. Anim. Husb.*, **9**: 393.
(44) Pope, A. L., Moir, R. J., Somers, M. and Underwood, E. J. (1968). *J. Anim. Sci.*, **27**: 1771.
(45) Reis, P. J. (1969). *Aust. J. Biol. Sci.*, **22**: 745.
(46) Roberts, R. B., Cowie, D. B., Abelson, P. H., Bolton, E. T. and Britten, R. J. (1955). *Studies of Biosynthesis in Escherishia coli.* Carnegie Institution, Washington. Publication 607.
(47) Salsbury, R. L., Marvil, D. K., Woodmansee, C. W. and Haenlein, G. F. (1971). *J. Dairy Sci.*, **54**: 390.
(48) Siebert, B. D. and Kennedy, P. M. (1972). *Aust. J. Agric. Res.*, **23**: 35.
(49) Sirotnak, F. M., Doetsch, R. N., Brown, R. E. and Shaw, J. C. (1953). *J. Dairy Sci.*, **36**: 1117.
(50) Starks, P. B., Hale, W. H., Garrigus, U. S., Forbes, R. M. and James, M. F. (1954). *J. Anim. Sci.*, **13**: 249.
(51) Thomas, W. E., Loosli, J. K., Williams, H. H. and Maynard, L. A. (1951). *J. Nutr.*, **43**: 515.
(52) Underwood, E. J. (1971). *Trace Elements in Human and Animal Nutrition.* Academic Press, New York.
(53) Virtanen, A. I., Ettala, T. and Makinen, S. (1972). *Proc. II Wld Congr. Anim. Feeding*, Madrid, **1**: 447.
(54) Whanger, P. D. and Matrone, G. (1970). Chapter 10 in Muth, O. H. (ed.), *Symposium: Sulphur in Nutrition.* Avi Publishing Co., Westport, Connecticut.
(55) Zikakis, J. P. and Salsbury, R. L. (1969). *J. Dairy Sci.*, **52**: 2014.

Effects of Sulphur Supplementation on Sheep and Cattle Production

A. M. DOWNES, J. P. LANGLANDS and P. J. REIS

The metabolism of S-containing compounds and their nutritional roles have been discussed in recent reviews (14,40). In Australasia, sheep and cattle are maintained almost exclusively at pasture and receive dietary S from forages and drinking water. S occurs in forages in both organic and inorganic forms and in drinking water mainly as sulphates. Most of the organic S is present as the sulphur-containing amino acids ('S-amino acids') cysteine, cystine, or methionine.

Irrespective of the chemical form, most of the S ingested by the ruminant is reduced to sulphide by the rumen microbes and may then be either incorporated into microbial protein or absorbed and oxidized to sulphate in the liver (2). Thus, in contrast to monogastric species, a large part of the S-amino acids utilized by ruminants is not provided directly by the diet but is derived from rumen microbes. Two types of S-supplementation may therefore be distinguished:

1. supplementation to maintain dietary S intake at a level at which microbial protein synthesis is not limited by the supply of S; and
2. the provision of additional S-amino acids in such a way that microbial degradation in the rumen is avoided.

This review deals with identifying and correcting S deficiencies, the two types of supplementation, problems due to excessive S intake and practical aspects of providing supplements.

Low-S forages generally contain large amounts of lignin and fibre and small amounts of protein and soluble carbohydrate. Such forages are of low organic matter digestibility, and energy intake is likely to be small even when S-deficiencies are corrected. As a result, the response to S is generally restricted to either a reduction in the rate of liveweight loss or a small gain in liveweight.

When the S content of the diet is reduced there is a depression in the concentration of sulphide in the rumen liquor and of sulphate in serum (28) and saliva (9). Ruminal microbial activity is also inhibited, and this has several undesirable consequences.

1. Microbial protein synthesis in the rumen is reduced (27); as a result, the utilization of dietary and particularly urea N is reduced, and the

concentrations of urea in blood and saliva and the excretion of N in urine are increased (9). Nitrogen retention therefore declines.

2. Less organic matter is digested in the rumen, and this results in an overall reduction in the animal's ability to digest dietary organic matter. Energy retention is therefore depressed.
3. Intake is reduced (12).
4. The conversion of ruminal lactate to propionate through the acrylate pathway is inhibited, and lactate accumulates in the rumen (61) if the diet contains large quantities of soluble carbohydrate.

Techniques are not available for accurately assessing either S intake or S status of grazing animals. However, some indication of the S content of the diet selected can be obtained from blood and faecal analyses. Serum sulphate-S was shown by Kennedy and Siebert (28) to be related to S intake. They found serum sulphate-S values of 8 μg/ml in heifers grazing poor quality tropical pasture in the dry season; this increased to 45 μg/ml when green herbage, which presumably contained adequate S, became available after rain. Faecal analysis has been suggested as a means of estimating the N and P contents of the diet selected by grazing cattle (36,37). Analyses of approximately 200 S balance experiments, in which sheep were given forages (31), suggested a curvilinear relationship between forage and faecal S contents. A faecal S concentration of 0.15 g S/100 g organic matter corresponded to a forage S content of 0.1 g S/100 g organic matter. The relationship was imprecise (S.E. of estimate $\pm$ 0.06 g S), and probably varies between forage species and seasons of the year; thus faecal S concentration may only be taken as an approximate indication of dietary S content.

SUPPLEMENTATION TO CORRECT DIETARY S DEFICIENCIES

Many workers (e.g. 1,56) have supplemented S-deficient diets under pen feeding conditions with inorganic S and have reported increased liveweight gains and, in some cases, increased wool production. Responses are seldom obtained when the diet contains more than 0.1 per cent S. For example, Coombe *et al.* (12) gave penned Merino wethers a diet of oaten straw and urea containing 0.06 per cent S and 1.3 per cent N. They supplemented this diet with Na, K, Ca, Mg, P, and S in various combinations. Animal performance was increased only when supplementary S was given. Feed intake by S-supplemented sheep was greater than that of the other groups, organic matter digestibility was higher, liveweight losses were reduced and wool production was increased.

Grazing animals have been given inorganic S as a supplement in a number of studies (e.g. 53,57). However, in contrast to results with penned animals, no convincing responses in production were observed and we are unaware of any reports of positive responses in more recent literature. Two reasons for the lack of response are suggested. Firstly, an essential nutrient other than S, such as N or P, might also have been limiting; and secondly, the forage actually selected by the animals may have contained adequate S even though chemical analysis of the forage available may have indicated an S-deficiency. Sheep and cattle select a diet higher in N, and presumably in S, than the average forage on offer. Sheep graze more selectively than cattle grazing the same pasture (Langlands, unpublished), but the digestive processes in cattle appear to be less sensitive than those of sheep to diets of low S content (28). This is possibly a result of less S being available for recycling in sheep than in cattle due to the proportionately higher requirement of S for wool growth than for hair growth.

A difficulty in identifying an S-deficiency in many experiments is that the effects of S and other supplements have been confounded and it is impossible to attribute unequivocally any response in production to the S component. For example, a supplement of urea and molasses is frequently given. However, molasses contains both energy and up to 1 per cent S (54), and a response to this supplement could be attributable to energy, N, S, or some other component, or to an interaction between these components.

No information is available on the effect of an S supplement on the composition of the diet selected. However, if supplementation increased intake, available forage could decline in quantity and this might reduce the opportunities for selective grazing, with a corresponding decline in the N and S content of the forage eaten. In the long term, compensatory growth may also erode the beneficial effects of giving supplements to grazing animals. Compensatory growth is the ability of animals to compensate for low growth rate during a period of malnutrition by making greater liveweight gains in a subsequent period when ample feed is available.

SUPPLEMENTATION TO INCREASE THE QUANTITY OF S-AMINO ACIDS AVAILABLE TO THE TISSUES

Dietary supplements of S-amino acids are usually ineffective in increasing the quantity of S-amino acid absorbed unless the diet is deficient in sulphur (see previous section) because S-amino acids are degraded by rumen microbes. Thus L-cystine (20,32) and DL-methionine (11) did not stimulate wool growth when added to the diet. Although Bird (6) found that methionine was degraded relatively slowly in the rumen, Bird and Moir (8) observed only a small response in wool growth to ruminal infusions of methionine. Studies of plasma amino acid patterns and the fate of [^{35}S] amino acids included in the diet also indicated poor utilization of these amino acids (16,49).

Although only the L-isomers of amino acids are utilized directly by mammals for protein synthesis, many D-isomers can be converted to the L-form. Thus, D-methionine and DL-2-hydroxy-4-methylthiobutyrate (MHA) can be converted to L-methionine (35). Also, the S of methionine can be converted to L-cysteine-S by transulphuration (21). Cysteine and cystine are readily interconvertible by oxidation and reduction or through thiol-disulphide exchange.

Wool growth

Clean wool is a complex mixture of proteins, some of which have a high content of cystine, and hence S. The S content of wool varies over a wide range; for example Reis (42) found values from 2.7 to 4.2 per cent S (9.8 to 15.3 per cent cystine) on a clean dry basis. These variations are due to altered proportions of certain components of the high-sulphur proteins which are extremely rich in cystine (24,25). The N:S ratio in wool is low (4 to 5) compared with the ratios for other animal proteins, ruminal microbes, and most forages (31,38,59).

Marston *et al.* (34) suggested that cystine was likely to be the primary nutrient limiting wool growth under pastoral conditions. The first direct evidence of the importance of cystine was provided by Marston (33) who recorded a response in wool growth in a sheep given L-cysteine subcutaneously. Subsequently Reis and Schinckel (45,46) found that the abomasal administration of either L-cysteine or DL-methionine caused a substantial increase in the rate of wool growth, and the effects of S-amino acids have now been investigated in some detail.

Sheep with a high genetic potential for wool growth respond to S-amino acid supplements to a greater extent than those with a low genetic potential (64). Diet may also be important. Most published experiments in which abomasal supplements of S-amino acids have stimulated wool growth, have been with penned sheep receiving roughage diets or with grazing sheep. Some recent experiments have indicated that the response to methionine is dependent on diet. Hemsley and Ferguson (unpublished) observed that intraperitoneal infusions of DL-methionine were ineffective in stimulating wool growth in sheep receiving a diet of wheat grain and wheaten hay (3:1). Further, abomasal infusions of methionine as a supplement for a diet consisting solely of wheat caused the production of very weak fibres with a reduced diameter (Reis, unpublished).

The amount of additional cyst(e)ine or methionine required for the maximal stimulation of wool growth has been investigated for sheep receiving 800 g/day of a diet of lucerne and wheaten hays (1:1). With this diet, 1.0 to 1.5 g/day S-amino acids would be available for absorption from the intestines (43). The optimal amount of supplementary methionine on this diet was 1 to 2 g/day when given as an abomasal infusion (43,48). Smaller amounts of methionine (0.5 to 0.6 g/day) provided sub-maximal responses (43,48), whereas greater quantities (6 to 10 g/day) had an adverse effect and sometimes depressed wool growth (43,44,48). The effects of cyst(e)ine have not been studied in detail, but there does not appear to be a sharply defined optimum dose, because large amounts (up to 8 g/day) did not reduce wool growth rate (43,48).

At low levels of supplementation (1 to 2 g/day) cystine, cysteine and methionine produce quantitatively similar responses in wool growth (43,48). A direct comparison (64) of abomasal supplements of equimolar amounts of L-cystine (2 g) or DL-methionine (2.5 g) indicated a slightly greater response to methionine. However, a detailed study of the dose-response curve for each amino acid would be needed to evaluate this result. The basis for the effectiveness of methionine has not been established. It could be due solely to its conversion to cysteine either in organs such as the liver (35) or in the skin or wool follicles (18), or it could have specific effects as suggested by Reis (43). MHA is also effective as an abomasal supplement for wool growth (43,44) but no direct comparisons have been made with methionine. It is not known whether the S of MHA is converted directly to cysteine-S as distinct from supplying methionine to the sheep.

Various parenteral routes of administration are effective for giving supplements of L-cysteine (17,19), L-methionine (17), and DL-methionine (4,5,29, 63). Direct comparisons with L-cyst(e)ine and L-methionine have demonstrated that the abomasal, duodenal or intravenous routes are equally effective for supplying S-amino acids for wool growth (3,48). S-amino acid supplements have usually been given as infusions for 6 to 24 hours each day. However, single daily injections of methionine, or even less frequent dosing, also stimulate wool growth (4,29,50,63).

Growth and milk production

Growth and milk production differ from wool growth in two important respects in relation to responses to S-amino acids. Tissue and milk proteins are not rich in S-amino acids, and there is a high energy requirement associated with their synthesis. Growth and milk production are usually restricted by energy intake and the proportion of calories supplied as protein rather than by a deficiency of S-amino acids.

The role of methionine as an essential amino acid for growth in young animals is well established, and it is frequently the first limiting amino acid. Walker and Kirk (60) demonstrated that an adequate supply of methionine plus cystine was essential for non-ruminant lambs if high rates of nitrogen retention were to be obtained, and abomasal infusions of methionine have increased nitrogen retention (8,50,51,52) and body-weight gains (43, 45) of adult sheep receiving restricted intakes of energy and N.

In ruminants producing large amounts of milk the output of S-amino acids in milk proteins is relatively greater than the output of cystine in wool by wool-producing sheep. Thus, a high-producing sheep may produce 15 g clean dry wool/day, which would contain 1.5 to 2.3 g cystine, whereas a high-producing goat (similar in size to a sheep) could be expected to secrete milk containing 3 to 4 g/day S-amino acids. However, while there is some evidence that methionine may be the first limiting amino acid for milk protein synthesis in dairy cows (10), the supply of this amino acid to the udder is usually adequate (23,55).

PROBLEMS ASSOCIATED WITH EXCESSIVE S CONSUMPTION

Excessive quantities of dietary S depress intake and in extreme cases can be fatal (62). The generation of large quantities of hydrogen sulphide in the rumen depresses rumen motility, and causes severe nervous and respiratory distress if it is absorbed through the lungs during eructation (15). Bird (7) proposed that the concentration of added S as sulphate, cystine or methionine should not exceed 0.2 per cent of the dry matter and suggested that this concentration would contain enough S for microbial protein synthesis without depressing intake. S supplementation may also induce Cu deficiency in the ruminant (13). Metabolic processes involving S, Cu, and Mo interact (58) and when S and Mo intakes are both increased, the absorption and retention of Cu are reduced, Cu storage in liver and blood declines, and a number of other manifestations of Cu deficiency including the production of 'steely' wool may be observed.

Deleterious effects are also observed when excessive quantities of methionine are absorbed. Daily abomasal infusions of 6 to 10 g methionine result in the accumulation of large amounts of methionine in plasma (47,49) and frequently depress wool growth (43,44,48).

THE PROVISION OF S SUPPLEMENTS

The evidence suggests that worthwhile responses in production would be obtained by giving inorganic S supplements to sheep and cattle when the diet contains less than 0.1 per cent S, and by giving S-amino acids to sheep to increase wool growth providing ruminal degradation of these compounds can be prevented.

Dietary deficiencies of S are likely to be associated with N deficiencies under grazing conditions, and supplements of urea, molasses and sufficient inorganic S to give an N:S ratio of approximately 10, provide a low-cost and effective supplement. Such supplements are widely used in commercial practice, and may be given in liquid form in a roller drum (39), or as a dry lick. Protein meals are more expensive, but there is less risk of toxicity and they contain both N and S. Two problems are associated with this form of supplementation under grazing conditions. Firstly, it is difficult to establish that a supplement is required; and secondly, it is difficult to regulate intake so that all animals receive the required quantity. Both problems result in inefficient use of supplement.

Although it is now established that wool growth responds to S-amino acids there are major difficulties in increasing the quantity absorbed by grazing sheep. One possible method of providing S-amino acid supplements in practice is to allow the supplement to enter the rumen but to protect the S-amino acids from microbial attack with subsequent release of the compound in the abomasum or intestines. Ferguson *et al.* (22) have summarized some methods of increasing the quantity of protein reaching the intestines, and the same principles apply to amino acids. They include reducing the time spent by amino acids in the rumen, reducing the activity of rumen microbes responsible for degrading S-amino acids, modifying the physical or chemical nature of the supplement, and applying a protective coating. A number of these methods have been examined. Neudoerffer, Duncan and Horney (41) reported that 60 per cent of orally administered methionine incorporated in a kaolin-fat matrix became available for post-ruminal absorption in cattle. MHA is less soluble than methionine and may be less readily degraded in the rumen. Griel *et al.* (26) found dietary MHA stimulated milk production by dairy cows. The compound has been investigated as a dietary supplement for wool growth. Reis (44) and Wickham (63) observed that MHA was ineffective, whereas Langlands (30) reported small increases in wool growth, but these were substantially less than the response to MHA given directly into the abomasum.

The problem of ruminal degradation of amino acids may be avoided by parenteral administration, and intraperitoneal injection can be used to establish a depot within the sheep from which S-amino acids are slowly released. This method is attractive for application to grazing animals because the quantity of S-amino acids for each animal can be directly controlled, and if the amino acid is released at an appropriate rate, it should only be necessary to muster the sheep infrequently. Sheep will tolerate intraperitoneal injections of at least 100 g of either cystine or methionine (17,29). Methionine, being rapidly absorbed and excreted, is ineffective but L-cystine is much less soluble and is released from the peritoneal cavity at a rate sufficient to increase wool growth over several weeks. Under grazing conditions, wool growth has been increased by 20 to 25 per cent by 28 g L-cystine given as an intraperitoneal implant at 28-day intervals (29). Difficulties with this procedure include the risk of infection associated with intraperitoneal injections under commercial conditions, and the formation of fatty adhesions around the implant which impede the absorption of cystine. Clearly more work is needed on this and other techniques of supplementing sheep with S-amino acids so that recent knowledge of the S requirements for wool growth can be applied commercially.

REFERENCES

(1) Albert, W. W., Garrigus, U. S., Forbes, R. M. and Norton, H. W. (1956). *J. Anim. Sci.*, **15**: 559.
(2) Anderson, C. M. (1956). *N.Z. J. Sci. Technol. A.*, **37**: 379.
(3) Barger, I. A., Southcott, W. H. and Williams, V. J. (1973). *Aust. J. Exp. Agric. Anim. Husb.*, **13**: 351.
(4) Barry, T. N. (1971). *Proc. N.Z. Soc. Anim. Prod.*, **31**: 129.
(5) Barry, T. N., Fennessy, P. F. and Duncan, S. J. (1973). *N.Z. J. Agric. Res.*, **16**: 64.
(6) Bird, P. R. (1972). *Aust. J. Biol. Sci.*, **25**: 185.
(7) Bird, P. R. (1972). *Aust. J. Biol. Sci.*, **25**: 1087.
(8) Bird, P. R. and Moir, R. J. (1972). *Aust. J. Biol. Sci.*, **25**: 835.
(9) Bray, A. C. and Hemsley, J. A. (1969). *Aust. J. Agric. Res.*, **20**: 759.
(10) Brown, R. E. (1969). In Swan, H. and Lewis, D. (eds), *Proceedings 3rd Nutrition Conference for Feed Manufacturers, University of Nottingham.* J. & A. Churchill Ltd, London, p. 23.
(11) Colebrook, W. F., Ferguson, K. A., Hemsley, J. A., Hogan, J. P., Reis, P. J. and Weston, R. H. (1968). *Proc. Aust. Soc. Anim. Prod.*, **7**: 397.
(12) Coombe, J. B., Christian, K. R. and Holgate, M. D. (1971). *J. Agric. Sci.* (Camb.), **77**: 159.
(13) Dick, A. T. (1953). *Aust. Vet. J.*, **29**: 233.
(14) Dodgson, K. S. and Rose, F. A. (1966). *Nutr. Abstr. Rev.*, **36**: 327.
(15) Dougherty, R. W., Mullenax, C. H. and Allison, M. J. (1965). In Dougherty, R. W. *et al.* (eds), *Physiology of Digestion in the Ruminant.* Butterworths, Washington D.C., p. 159.
(16) Downes, A. M., Reis, P. J., Sharry, L. F. and Tunks, D. A. (1970). *Br. J. Nutr.*, **24**: 1083.
(17) Downes, A. M., Reis, P. J., Sharry, L. F. and Tunks, D. A. (1970). *Aust. J. Biol. Sci.*, **23**: 1077.
(18) Downes, A. M., Sharry, L. F. and Till, A. R. (1964). *Aust. J. Biol. Sci.*, **17**: 945.
(19) Dryden, G. M., Wickham, G. A. and Cockrem, F. (1969). *N.Z. J. Agric. Res.*, **12**: 580.
(20) Du Toit, P. J., Malan, A. I., Groenewald, J. W. and Botha, M. L. (1935). *Onderstepoort J. Vet. Sci. Anim. Ind.*, **4**: 229.
(21) Du Vigneaud, V. (1952). *A Trail of Research in Sulfur Chemistry and Metabolism.* Cornell University Press, New York.
(22) Ferguson, K. A., Hemsley, J. A. and Reis, P. J. (1967). *Aust. J. Sci.*, **30**: 215.
(23) Fisher, L. J. (1969). *J. Dairy Sci.*, **52**: 943.
(24) Gillespie, J. M., Broad, A. and Reis, P. J. (1969). *Biochem. J.*, **112**: 41.
(25) Gillespie, J. M. and Reis, P. J. (1966). *Biochem. J.*, **98**: 669.
(26) Griel, L. C., Patton, R. A., McCarthy, R. D. and Chandler, P. T. (1968). *J. Dairy Sci.*, **51**: 1866.
(27) Hume, I. D. and Bird, P. R. (1970). *Aust. J. Agric. Res.*, **21**: 315.

(28) Kennedy, P. M. and Siebert, B. D. (1972). *Aust. J. Agric. Res.*, **23**: 45.
(29) Langlands, J. P. (1970). *Aust. J. Exp. Agric. Anim. Husb.*, **10**: 665.
(30) Langlands, J. P. (1972). *Proc. Aust. Soc. Anim. Prod.*, **9**: 321.
(31) Langlands, J. P., Sutherland, H. A. M. and Playne, M. J. (1973). *Brit. J. Nutr.*, **30**: 537.
(32) Marston, H. R. (1932). *Aust. J. Exp. Biol. Med. Sci.*, **9**: 235.
(33) Marston, H. R. (1935). *J. Agric. Sci.* (Camb.), **25**: 113.
(34) Marston, H. R. and Brailsford Robertson, T. (1928). Bull. Coun. Scient. Ind. Res., Melbourne No. 39.
(35) Meister, A. (1965). *Biochemistry of the Amino Acids* Vol. 2, 2nd edn. Academic Press, New York and London.
(36) Moir, K. W. (1960) *Qd J. Agric. Sci.*, **17**: 361.
(37) Moir, K. W. (1960). *Qd J. Agric. Sci.*, **17**: 373.
(38) Moir, R. J., Somers, M. and Bray, A. C. (1967-8). *Sulphur Inst. J.*, **3**: 15.
(39) Morris, J. G. and Gulbransen, B. (1970). *Aust. J. Exp. Agric. Anim. Husb.*, **10**: 379.
(40) Muth, O. H. and Oldfield, J. E. (1970) (eds). *Symposium: Sulphur in Nutrition.* Avi Publishing Co., Westport, Connecticut.
(41) Neudoerffer, T. S., Duncan, D. B. and Horney, F. D. (1971). *Brit. J. Nutr.*, **25**: 333.
(42) Reis, P. J. (1965). In Lyne, A. G. and Short, B. F. (eds), *Biology of the Skin and Hair Growth.* Angus and Robertson, Sydney, p. 365.
(43) Reis, P. J. (1967). *Aust. J. Biol. Sci.*, **20**: 809.
(44) Reis, P. J. (1970). *Aust. J. Biol. Sci.*, **23**: 441.
(45) Reis, P. J. and Schinckel, P. G. (1963). *Aust. J. Biol. Sci.*, **16**: 218.
(46) Reis, P. J. and Schinckel, P. G. (1964). *Aust. J. Biol. Sci.*, **17**: 532.
(47) Reis, P. J. and Tunks, D. A. (1971). *Search*, **2**: 108.
(48) Reis, P. J., Tunks, D. A. and Downes, A. M. (1973). *Aust. J. Biol. Sci.*, **26**: 249.
(49) Reis, P. J., Tunks, D. A. and Sharry, L. F. (1973). *Aust. J. Biol. Sci.*, **26**: 635.
(50) Robards, G. E. (1971). *Aust. J. Agric. Res.*, **22**: 261.
(51) Schelling, G. T. (1970). *Fedn Proc. Fedn Am. Socs Exp. Biol.*, **29**: 759.
(52) Scott, R. A., Little, C. O., Amos, H. E. and Mitchell, G. E. (1972). *J. Anim. Sci.*, **35**: 446.
(53) Seddon, H. R. and Chamberlin, W. E. (1933). *Aust. Vet. J.*, **9**: 154.
(54) Siebert, B. D. and Kennedy, P. M. (1972). *Aust. J. Agric. Res.*, **23**: 35.
(55) Teichman, R., Caruolo, E. V. and Mochrie, R. D. (1969). *J. Dairy Sci.*, **52**: 942.
(56) Thomas, W. E., Loosli, J. K., Williams, H. H. and Maynard, L. A. (1951). *J. Nutr.*, **43**: 515.
(57) Underwood, E. J. (1934). *J. Dep. Agric. West. Aust.*, **11**: 507.
(58) Underwood, E. J. (1966). *Mineral Nutrition of Livestock.* FAO and Commonwealth Agricultural Bureaux, Farnham Royal, Bucks.
(59) Walker, D. J. and Nader, C. J. (1968). *Appl. Microbiol.*, **16**: 1124.
(60) Walker, D. M. and Kirk, R. D. (1972). *Proc. II Wld Congr. Anim. Feeding,* Madrid, October 1972, p. 337.
(61) Whanger, P. D. and Matrone, G. (1967). *Biochim. Biophys. Acta,* **136**: 27.
(62) White, J. B. (1964). *Vet. Rec.*, **76**: 278.
(63) Wickham, G. A. (1970). *Proc. N.Z. Soc. Anim. Prod.*, **30**: 209.
(64) Williams, A. J., Robards, G. E. and Saville, D. G. (1972). *Aust. J. Biol. Sci.*, **25**: 1269.

4 The Incidence and Correction of Sulphur Problems in Australasia

The Occurrence of Sulphur Deficiency in the Australian Tropics

R. K. JONES, M. E. PROBERT and B. J. CRACK

Before the use of inorganic fertilizers and pollution of the atmosphere by industrial activity, non-nodulated plants obtained most of their N and S requirements from the breakdown of organic matter in the soil. The ratio of N:S in soil is usually between 6 and 10 (Table 4-1). In the short term N and S may be mineralized in varying ratios, but in the long term they appear to be mineralized in about the same ratio as in the organic matter (54). The N:S ratio in healthy plants is much higher, and values between 13 and 18 are generally found (13). It follows that N will limit plant growth before S if soil organic matter is the main source of these nutrients. If nitrogen is added to the system as fertilizer or by a legume, then there is a greater chance of S deficiency developing.

LAND USE

For the purposes of this paper the 'Australian tropics' are defined as the whole of the state of Queensland plus those portions of Western Australia and the Northern Territory north of the Tropic of Capricorn. There are relatively few published reports on S which have originated in this enormous area. In 1970, Williams and Andrew (56) were unable to outline any major occurrences of S deficiency in the area defined. It cannot be inferred, however, that S is adequate on all soils and one must look further for an explanation of this apparent sufficiency of S.

A large proportion of the area is under native grassland which is grazed extensively. Under this system, little if any attempt is made to overcome the widespread deficiencies of N and P (40), so one would probably not look for, or expect to find, responses to S. Intensive agriculture based on crops such as sugar cane, bananas and pineapples is restricted to small pockets receiving a high rainfall or where irrigation is available. Sugar cane is heavily fertilized; in 1970-1 an average of 766 kg of fertilizer was applied per ha. However, even without fertilizers, Sedl (36) has shown that sufficient S is available to the crop for maximum yield. Between these two extremes of land use there are considerable areas used for improved pastures or cropping or with potential for such developments. Here, environmental and other factors make it feasible to introduce legumes and to at least partially overcome deficiencies of other nutrients such as N, P, K, and Mo. It is under these conditions that one would expect to find a greater incidence of S deficiency.

At present, fertilizer use and hence S input on pastures and grain crops in

TABLE 4-1
Mean C:N:S ratios and sulphur contents of surface soil samples

Location and/or	Soil type	C:N:S	Per cent S	Reference
North Queensland	Solodics	133:10:1.3	0.010	(10)
	Neutral red duplex	148:10:1.3	0.009	(11)
	Granitic sands	197:10:—	0.004	(22)
	Red earths	200:10:1.1	0.006	CSIRO, Division of Soils, Townsville (unpublished)
	Yellow earths	221:10:1.2	0.007	
	Krasnozems	143:10:1.7	0.075	
	Euchrozems	171:10:1.2	0.017	
	Mareeba-Dimbula area (various)	—:10:1.1	0.004	(32)
Eastern Australia	agricultural	150:10:1.26		(57)
Australia	grasslands	113:10:1.2		(16)
New Zealand	grasslands	130:10:1.3	0.028–0.091	(50)
England and Wales	(mineral soils)	100:10:1.8	0.048	(23)
Scotland		140:10:1.38		(59)
Iowa		109:10:1.5	0.029	(42)
West Indies		125:10:1.1	0.025	(20)
Australia	calcareous		0.009 –0.198	(58)
	acid		0.0025–0.055	(58)
East Africa	forest top soils		0.023 –0.075	(21)

the Australian tropics is at a very low level. Although superphosphate usage is increasing steadily (Figure 4-1), Queensland and the Northern Territory together used less than three per cent of the Australian total of 3.7×10^6 tonnes in 1970-1. On average, pastures which were fertilized in 1970-1 received 204 kg/ha of superphosphate, but the area fertilized (155,000 ha) was only 5.5 per cent of the area of effective sown pasture (2) and a negligible percentage of the potential area for pastures (12). Similarly, only 25 per cent of the total crop area received fertilizer in 1970-1—the average for the other states was 67 per cent.

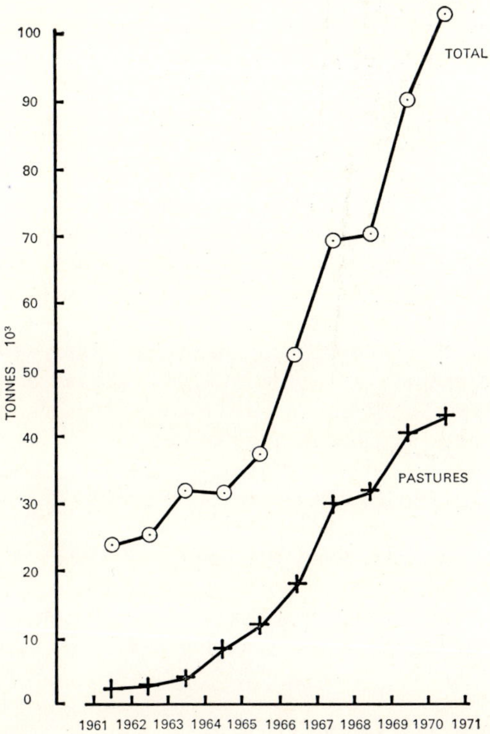

Figure 4-1 Usage of superphosphate in Queensland and the Northern Territory between 1961 and 1971.
(Data obtained from Commonwealth Bureau of Census and Statistics.)

SULPHUR IN SOIL

The N:S ratios of various groups of soils from the Australian tropics are similar to those reported from other parts of the world (Table 4-1). Several of the tropical soils, however, are high in carbon relative to N and S and one would expect this to be unfavourable to mineralization of S (6). Apart from the soils developed on basalt, those from tropical Australia are very low in total S. Nevertheless they are still within the range for Australian soils reported by other workers (Table 4-1).

In general, sulphur in surface soils is contained largely in the organic matter fraction where it is unavailable to plants unless mineralized. Total S and organic S in the soil have been found to be poor indices of plant response to applied S (17) nor are they well related to the rate of S mineralization (41). Better relationships have been found between plant uptake of S and various 'extractable soil S values', particularly phosphate-extractable S (7,23). This topic is the subject of other papers at the symposium.

In tropical Australia S mineralization should not be limited by the temperature restraints reported in other environments (55). However, on the Darling Downs in Queensland low temperatures do occur and Jones (28) interpreted the seasonal nature of his S responses to such an effect. Over much of the region the marked seasonal distribution of rainfall would probably result in greater moisture restraints to mineralization than do low temperatures.

The accession of S by the plant/soil system in this non-industrialized area, if indeed it can be considered a gain rather than a redistribution (52), would be expected to be low. At Townsville and Woodstock (20 km inland), depositions of 5.7 and 2.6 kg S/ha/annum respectively have been measured (Probert, unpublished). The only other records available are from five coastal stations in Queensland (36) where annual depositions of 3.7 to 25.2 kg S/ha were measured. The higher figures were registered during periods of intense rainfall associated with cyclones.

Burning is likely to be important in the redistribution of S in tropical Australia. Most native pastures are subject to fires (48) but we have been unable to find any reports on the losses of S from such fires. The losses are likely to be appreciable if data from the burning of heather (36 per cent loss) are any guide (14). It seems probable that some of the S in rainfall determined by Sedl (36) during the July to December period originates from the 'firing' of cane trash.

Sulphur can also be lost from the soil/plant system by erosion and by leaching. An extreme example of leaching was reported by Gillman (18) who found that the loss of sulphur, applied as superphosphate, from a 100 cm profile of a siliceous sand in Cape York Peninsula, was virtually complete in one wet season.

PLANT RESPONSES TO SULPHUR

Table 4-2 summarizes all the published work showing moderate to strong responses to S in tropical Australia. Some of this work (e.g., 27,31) refers to soils of only minor importance, so is of little use in assessing the overall situation. Other work (e.g., 9,25) refers to the results of pot experiments only and it is dangerous to extrapolate to the field situation without other evidence. In recent unpublished work, for example, we found strong responses to S in pot experiments on some yellow earth soils at McDonnell near Cape York, but to date there is little evidence of a field response, presumably because of the increase in extractable S at depth in the profile (Table 4-3). The importance of considering the whole rooting depth of soil has been pointed out by Jones (28).

There is good evidence (Table 4-2) of sulphur deficiency on the granite sands in Cape York Peninsula (22,24), on the lateritic-podzolic soils of the

A summary of published work from tropical Australia in which responses to S have been recorded

Soil type and location	Type of expt and test plant	Data on S in soil or plant	Results	Reference
4 diverse soils from Katherine, N.T.	Pot (subterranean clover)	Nil	Moderate response on 2 soils, strong on others	(33)
Deep sandy soils on granite, Cape York Peninsula, Qld	Pot and field (Townsville stylo)	Total soil S for 23 surface samples and for 3 profiles to 180 cm. Total plant S for selected samples	Very strong responses to S in pots on 8 soils and in the one field expt	(22, 24)
8 soils mainly of light texture from Mareeba area, Qld	Pot (tobacco)	Total, HCO_3 and PO_4 extractable S on 0–30 cm soil samples. Total S on plant leaves	Moderate-strong responses to S on all soils	(32)
9 soils mainly of light texture from Mareeba area, Qld	Pot (tomato)	Nil	Moderate to strong responses on 4 virgin soils. No responses on cropped soils	(51)
Solodic soils in coastal region near Townsville, Qld	Pot (Phasey bean and Townsville stylo)	Total S profiles on 3 selected soils. Some total plant S values	Moderate to strong responses to S on all 7 soils	(10, 25)
Solodic soil, Woodstock near Townsville, Qld	Pot and field (Townsville stylo)	Total and sulphate S on plant tops, S:N ratios on total and on alcohol insoluble material	Moderate response to S in both pot and field. Soil not of widespread importance	(26, 27)
Neutral red duplex soils near Townsville, Qld	Pot (Phasey bean and Townsville stylo)	Total S, HCO_3 and PO_4 extractable S on 15 surface soils. Total S on 3 selected profiles. A few plant S values	Strong responses to S on all soils	(9, 11)
Solodic soil, Rodds Bay, Qld	Pot (Townsville stylo)	Nil	Moderate response to S in growth and N content	(37)
Lateritic-podzolic soil, Beerwah, Qld	Field (White clover)	Total S on plant tops	Strong responses to S when other deficiencies corrected	(4)
Solodic soils on granodiorite, Eskdale, Qld	Field (Rhodes grass and Siratro)	Nil	Moderate responses to S once P deficiency corrected	(34)
A shallow dark brown clay on basalt, eastern Darling Downs, Qld	Pot and field (lucerne)	Nil	Moderate pot and field responses in growth. Soil shallow and not of widespread importance	(31)
Alluvial clay soil near Warwick, Qld	Field (lucerne)	Total S on plant material at 5 harvests	Strong responses in yield, per cent S and per cent protein to various forms and rates of S	(30)
6 soils derived from basalt, Eastern Darling Downs, Qld	Field (lucerne)	Total soil S at 0–15 and 31–46 cm. Profiles of water soluble S on 4 soils	Moderate responses to S at some sites, mainly in winter-spring period	(28)
Black clay soil, Brookstead, Qld	Pot (Phasey bean, Setaria italica, and Paspalum scrobiculatum)	Total S and sulphate S on plant tops	Strong responses to S and N	(5)
Sandy solodic soil, Beaudesert, Qld	Field (White clover)	Nil	Moderate response to S, particularly in presence of Mo	(49)
2 soils on basic igneous rocks near Boonah, Qld	Pot (Phasey bean) and field (lucerne)	Total S on 2 soils (0–15 cm)	Moderate response on 2 soils in pots; responses in spring only at the one field site	(35)

TABLE 4-3
Phosphate extractable sulphur* in some North Queensland soil profiles (Probert, unpublished) (ppm S, oven dry basis)

Soil type	Soils derived from basalt			Yellow earths			Soils of heavy texture			
		Krasnozem	Euchrozem				Grey-brown clay		Grey clay (brigalow)	
PPF**		Gn 3.11	Gn 3.12		Gn 2.64	Gn 2.24		Ug 5.26		Ug 5.24
Site		Malanda	Boomerang		McDonnell	Yarrowmere		Hughenden		Bletchington Park
	Depth (cm)	S	S	Depth (cm)	S	S	Depth (cm)	S	Depth (cm)	S
	0– 10	40	5	0– 10	5	2	0–15	7	0– 10	3
	10– 20	62	3	10– 20	5	3	15–30	3	10– 20	13
	20– 30	61	2	20– 30	5	3	30–60	13	20– 30	29
	30– 60	107	2	30– 40	9	2	60–75	76	30– 60	59
	60– 90	453	1	40– 60	25	3	75–90	~4500	60– 90	38
	90–120	682	1	60– 90	31	3		(gypsum	90–120	13
	120–150	529	<1	90–120	61	2		present)	120–150	2
	150–180	721	<1	120–150	95	3			150–180	1
	180–210	711	1						180–200	2

* Method of Barrow (7).
** Of the Factual Key of Northcote (see (39)).

Wallum area near Brisbane (4), on several soils near Beaudesert (35,49), on some of the shallower black earths derived from basalt in the eastern Darling Downs (28,30), and in the Lockyer Valley (1).

Cameron (8) working on the cracking clays near Theodore failed to find consistent responses to S, although the clover component of the pasture gave occasional responses during spring.

In other work, little or no response to S has been found. Sugar cane gave no response in either yield or sugar content at nine sites along the Queensland coast, even after a succession of crops (36). On most of the soils phosphate-extractable S levels after cropping were moderate to high, particularly at depth; inputs of S in the rainfall were also appreciable. In a series of papers on soils in the Cardwell-Cairns area of Queensland, Teitzel and Bruce (43-47) found only occasional responses to S in experiments with pasture legumes.

Pot experiments on soils derived from basalt, acid volcanics, granites, and schists from the nearby Atherton Tableland also gave little evidence of S deficiency (29). The high extractable-S values for one of the basaltic soils (Malanda) shown in Table 4-3 suggest that field responses are also most unlikely. The situation on the Maleny Tableland is probably similar (53). In contrast, strong pot and field responses have been found on related soils under eucalypt woodland in drier areas (Miller and Jones, unpublished); this is consistent with the very low levels of extractable-S found throughout the profile—see Table 4-3, Boomerang.

Lack of response to S has also been documented in pot experiments on soils from under Brigalow (*Acacia harpophylla*) in central and southern Queensland (15), on two humic gley soils from the Wallum near Brisbane (3), and in a field experiment at Westwood (19).

SULPHUR AND ANIMAL NUTRITION

This topic is dealt with in greater detail in other papers. However, some interesting work has recently been done in tropical Australia by Siebert and Kennedy (38) who fed poor quality native pastures to sheep and cattle. They found significant responses in digestible energy intake from additions of S to the diet, provided that a favourable N:S ratio was maintained by the use of non-protein N. A larger proportion of S was recycled to the rumen of cattle than sheep.

AN OVERALL ASSESSMENT

It is clear that published information is insufficient to enable us to make much advance from Williams and Andrew (56) in delineating S responsive and non-responsive areas. However, many workers have kindly given us access to results from unpublished pot and field experiments. By collating these, along with the published results above, we are able to present a tentative map of the occurrence of S deficiency in Queensland (Figure 4-2). In essence the map refers to the probability of S deficiency limiting yields of legume based pastures, rather than more intensive forms of agriculture.

In all, the results of glasshouse experiments on 373 soils (see Figure 4-3 for distribution of sites) and of 80 field experiments and trials involving S as a treatment were available to us. Very little information was available

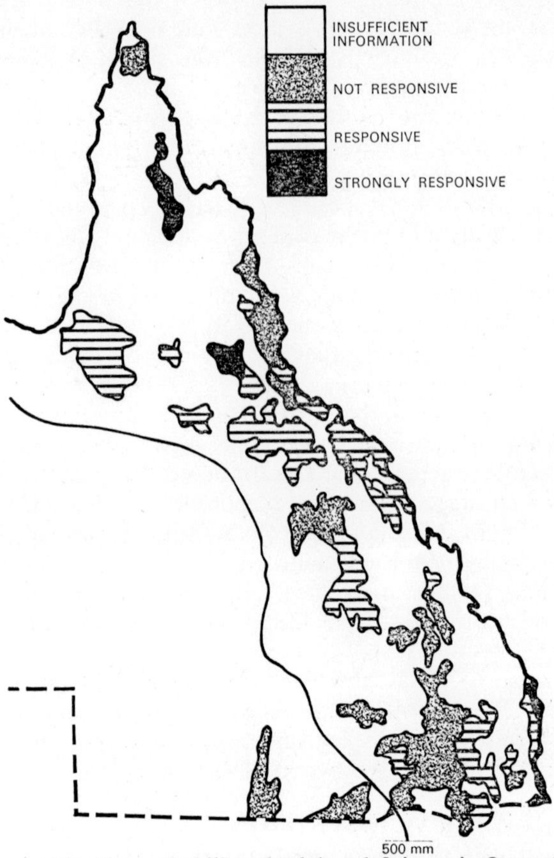

Figure 4-2 Map showing the probability of sulphur deficiency in Queensland. See text for further details on the various categories. The 500 mm rainfall isohyet is also shown.

from the north of Western Australia, the Northern Territory, and areas of Queensland with less than 500 mm of rainfall/annum, so we were unable to map these areas.

The basis of our extrapolation was the soil map in *A Handbook of Australian Soils* (39) which shows only the dominant soil groups. The danger of extrapolating S responses in pots to the field situation cannot be too strongly stressed. We have only used category 4, 'strongly responsive', in those situations where good responses have also been obtained in the field. The areas designated 'responsive' are those where pot experiments have given large responses (> 30 per cent yield response with added S), but modified according to field evidence or soil data. On soils where it is known that (a) extractable-S increases strongly with depth (see Table 4-3 for example), (b) gypsum nodules occur at depth, or (c) no response to S has been found in field experiments, we have used category 2 'not expected to be responsive', irrespective of the results of pot experiments on surface soils. Areas have also been classed in category 2 where a reasonable number of pot experiments on soils from the area has given no response. It is very important that the reader realize that

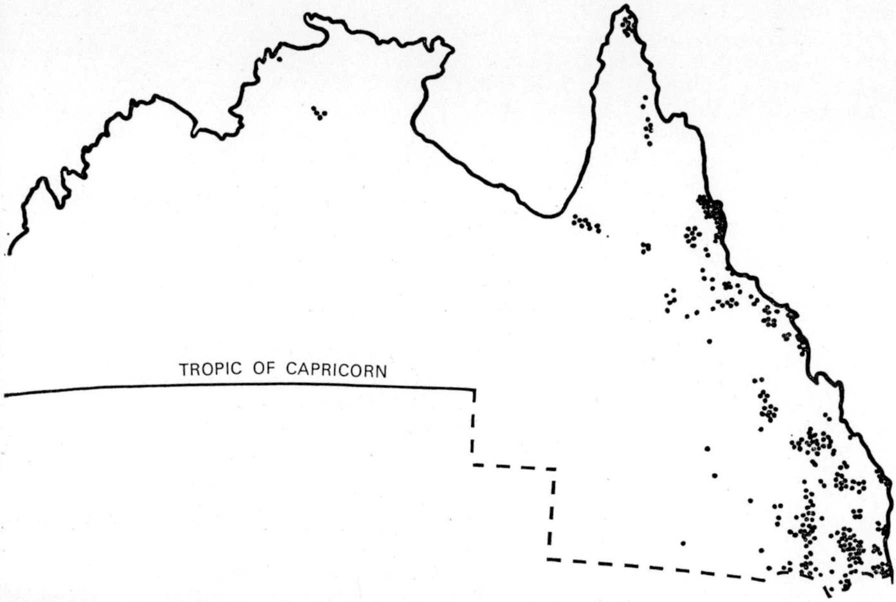

Figure 4-3 The location of soils in tropical Australia on which pot experiments involving S as a treatment have been carried out.

the map applies only to situations where deficiencies or toxicities of other nutrients have been corrected.

Despite the large number of pot and field experiments which have been carried out, the dominant feature of the resultant map is the large area for which there is little or no information. The tentative nature of the map must be stressed. Obviously, a great deal more work is required in order to characterize the sulphur status even at the level of detail attempted here, and this could well modify our classification.

Acknowledgements. We are very grateful to colleagues employed by the Queensland State Government, CSIRO, and others, who kindly made information available to us. In particular we are grateful to Mr D. G. Cameron, Department of Primary Industries, Queensland, and Mr R. F. Isbell, CSIRO Division of Soils, Townsville, for assistance in preparing this manuscript.

REFERENCES

(1) Anon. (1948). Qld Dept of Agriculture and Stock, Annual Report 1947-8.
(2) Anon. (1972). *Sown pastures and seed production in Queensland.* Qld D.P.I., August 1972.
(3) Andrew, C. S. and Bryan, W. W. (1955). *Aust. J. Agric. Res.*, **6**: 265.
(4) Andrew, C. S. and Bryan, W. W. (1958). *Aust. J. Agric. Res.*, **9**: 267.
(5) Andrew, C. S., Kipps, E. H. and Barford, H. (1952). *Aust. J. Agric. Res.*, **3**: 111.
(6) Barrow, N. J. (1960). *Aust. J. Agric. Res.*, **11**: 960.
(7) Barrow, N. J. (1967). *Soil Sci.*, **104**: 242.
(8) Cameron, D. G. (1969). *Qd J. Agric. Anim. Sci.*, **26**: 373.
(9) Crack, B. J. (1971). *Aust. J. Exp. Agric. Anim. Husb.*, **11**: 336.
(10) Crack, B. J. and Isbell, R. F. (1970). *Aust. J. Exp. Agric. Anim. Husb.*, **10**: 334.

(11) Crack, B. J. and Isbell, R. F. (1971). *Aust. J. Exp. Agric. Anim. Husb.*, **11**: 328.
(12) Davies, J. G. and Eyles, A. G. (1965). *J. Aust. Inst. Agric. Sci.*, **31**: 77.
(13) Dijkshoorn, W. and Van Wijk, A. L. (1967). *Pl. Soil*, **26**: 129.
(14) Evans, C. C. and Allen, S. E. (1971). *Oikos*, **22**: 149.
(15) Fergus, I. F. (1962). CSIRO Division of Soils. Divisional report 1/62. Adelaide. South Australia.
(16) Freney, J. R. (1960). *Aust. J. Biol. Sci.*, **13**: 387.
(17) Freney, J. R., Barrow, N. J. and Spencer, K. (1962). *Pl. Soil*, **17**: 295.
(18) Gillman, G. P. (1973). *Aust. J. Exp. Agric. Anim. Husb.* (in press).
(19) Hall, R. L. (1970). *Tropical Grasslands*, **4**: 77.
(20) Haque, I. and Walmsley, D. (1973). *Trop. Agric.* (Trinidad), **50**: 55.
(21) Hesse, P. R. (1957). *Pl. Soil*, **9**: 86.
(22) Isbell, R. F. and Gillman, G. P. (1973). *Aust. J. Exp. Agric. Anim. Husb.*, **13**: 81.
(23) Jones, L. H. P., Cowling, D. W. and Lockyer, D. R. (1972). *Soil Sci.*, **114**: 104.
(24) Jones, R. K. (1973). *Aust. J. Exp. Agric. Anim. Husb.*, **13**: 89.
(25) Jones, R. K. and Crack, B. J. (1970). *Aust. J. Exp. Agric. Anim. Husb.*, **10**: 342.
(26) Jones, R. K. and Robinson, P. J. (1970). *Proc. XI Int. Grassld Congr.*, p. 377.
(27) Jones, R. K., Robinson, P. J., Haydock, K. P. and Megarrity, R. G. (1971). *Aust. J. Agric. Res.*, **22**: 885.
(28) Jones, R. M. (1970). *Aust. J. Exp. Agric. Anim. Husb.*, **10**: 749.
(29) Kerridge, P. C., Andrew, C. S. and Murtha, G. G. (1972). *Aust. J. Exp. Agric. Anim. Husb.*, **12**: 618.
(30) Littler, J. W. and Price, M. J. (1967). *Qd J. Agric. Anim. Sci.*, **24**: 159.
(31) Lloyd, D. L. (1969). *Tropical Grasslands*, **3**: 99.
(32) Lovett, W. J. and Crockford, R. H. (1969). *Fld Stn Rec. Div. Pl. Ind. CSIRO (Aust.)*, **8**: 77.
(33) McLachlan, K. D. (1953). *J. Aust. Inst. Agric. Sci.*, **19**: 197.
(34) t'Mannetje, L. (1967). *Tropical Grasslands*, **1**: 9.
(35) Rees, M. C. (1972). *Tropical Grasslands*, **6**: 205.
(36) Sedl, J. M. (1971). Technical Communication 1971, No. 2. Bureau of Sugar Experiment Stations, Brisbane, Queensland.
(37) Shaw, N. H., Gates, C. T. and Wilson, J. R. (1966). *Aust. J. Exp. Agric. Anim. Husb.*, **6**: 150.
(38) Siebert, B. D. and Kennedy, P. M. (1972). *Aust. J. Agric. Res.*, **23**: 35.
(39) Stace, H. C. T. *et al.* (1968). *A Handbook of Australian Soils.* Rellim Technical Publications, Glenside, South Australia.
(40) Stephens, C. G. and Donald, C. M. (1958). *Adv. Agron.*, **10**: 167.
(41) Tabatabai, M. A. and Bremner, J. M. (1972). *Agron. J.*, **64**: 40.
(42) Tabatabai, M. A. and Bremner, J. M. (1972). *Soil Sci.*, **114**: 380.
(43) Teitzel, J. K. and Bruce, R. C. (1971). *Aust. J. Exp. Agric. Anim. Husb.*, **11**: 77.
(44) Teitzel, J. K. and Bruce, R. C. (1972). *Aust. J. Exp. Agric. Anim. Husb.*, **12**: 49.
(45) Teitzel, J. K. and Bruce, R. C. (1972). *Aust. J. Exp. Agric. Anim. Husb.*, **12**: 281.
(46) Tietzel, J. K. and Bruce, R. C. (1973). *Aust. J. Exp. Agric. Anim. Husb.*, **13**: 306.
(47) Teitzel, J. K. and Bruce, R. C. (1973). *Aust. J. Exp. Agric. Anim. Husb.*, **13**: 312.
(48) Tothill, J. C. (1971). *Tropical Grasslands*, **5**: 1.
(49) Truong, N. V., Andrew, C. S. and Skerman, P. J. (1967). *Aust. J. Exp. Agric. Anim. Husb.*, **7**: 232.
(50) Walker, T. W. and Adams, A. F. R. (1958). *Soil Sci.*, **85**: 307.
(51) Ward, D. K. (1967). *Qd J. Agric. Anim. Sci.*, **24**: 253.
(52) Wetselaar, R. and Hutton, J. T. (1963). *Aust. J. Agric. Res.*, **14**: 319.
(53) White, R. E. (1967). *Aust. J. Exp. Agric. Anim. Husb.*, **7**: 509.
(54) Williams, C. H. (1967). Intern. Soc. Soil Sci. Trans. Comm. II and IV, p. 93.
(55) Williams, C. H. (1967). *Pl. Soil*, **26**: 205.
(56) Williams, C. H. and Andrew, C. S. (1970). In Milton Moore, R. (ed.), *Australian Grasslands.* Australian National University Press, Canberra, pp. 321ff.
(57) Williams, C. H. and Steinbergs, A. (1958). *Aust. J. Agric. Res.*, **9**: 483.
(58) Williams, C. H. and Steinbergs, A. (1962). *Pl. Soil*, **17**: 279.
(59) Williams, C. H., Williams, E. G. and Scott, M. N. (1960). *J. Soil Sci.*, **11**: 334.

The Occurrence of Sulphur Deficiency in Temperate Australia

G. J. BLAIR and A. J. NICOLSON

As in other countries the early recognition of sulphur deficiency in Australia was masked by the use of superphosphate. However, the earliest recorded experiments with superphosphate included gypsum treatments to which no responses were obtained (27).

Letters from the Glen Innes Research Station (New South Wales) in 1915 and 1926 mention responses to potassium sulphate and to gypsum respectively, and these responsive soils have since been shown to be adequately supplied with potassium and calcium. Glynne (12) recorded a nine-fold sulphur response in wheat to the application of 1.5 g S/kg soil, but this has subsequently been shown to be due to the effects of sulphur on manganese availability (18). Despite the isolated cases listed above, positive recognition of the importance of sulphur in Australia was delayed until the 1950s and '60s.

Chronologically, Anderson and Spencer (1) gave the first published experimental evidence of sulphur deficiency in Australia, reporting responses in pot culture on three soils from the Southern Tablelands of N.S.W. Later, McLachlan (21) obtained responses to sulphur, applied as sulphate on a basaltic soil of adequate phosphorus status, and Rossiter (28) reported the first documented account of sulphur responses in Western Australia. Sandy soils from Perth, Gingin and Kojonup, were shown responsive.

In 1963, Powrie (26), investigating the residual value of superphosphate on solodized solonetz soils near Keith, demonstrated responses to gypsum applications where the residual phosphorus status was high. Similar results were observed on lateritic podzolic soils near Mount Compass.

In 1967, Drake and Curnow (8,9) demonstrated that large areas of East Gippsland (Victoria) were sulphur deficient, particularly the coarser textured soils tested.

In 1964, at the Australian Plant Nutrition Conference, Paton reported responses to sulphur in pastures on acid sandy soils on Flinders Island, and the northeast coast of Tasmania and on peaty sands on the northwest coast. In 1969, Banath (2) obtained responses in subterranean clover to sulphur applied as pyrites, on a sandy soil from northeastern Tasmania.

SOILS RESPONSIVE TO SULPHUR

By the 1970s it was apparent that soils in many parts of temperate Australia were responsive to sulphur (Figure 4-4). Sulphur deficiency is now known to

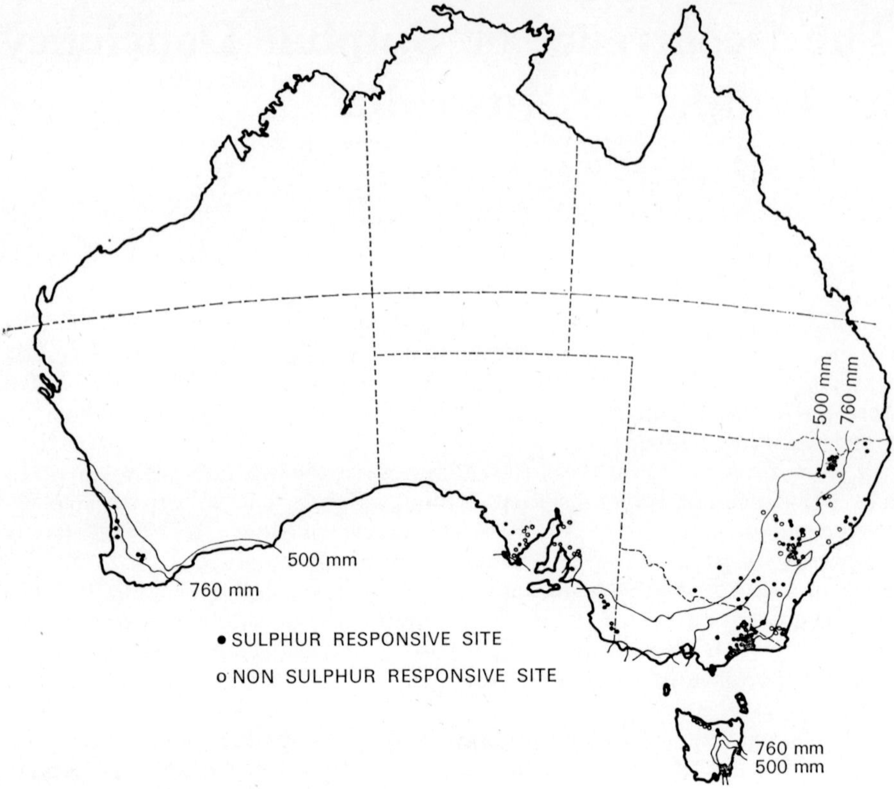

Figure 4-4 Sulphur test sites of temperate Australia.

limit growth in widespread areas of the Tablelands of New South Wales, the highlands of Gippsland in Victoria and areas of Northern Tasmania. Small areas of deficiency occur on sandy soils in South Australia, and large areas on the sandy soils in southwestern Western Australia. In many cases sulphur deficiency is accompanied by phosphorus deficiency and frequently the application of superphosphate provides adequate sulphur fertilization.

New South Wales

Seventy per cent of the soils of the Northern Tablelands of New South Wales ($>$ 4 million ha) have been estimated as moderately to very deficient in sulphur, and 80 per cent of the soils in the same area were similarly deficient in phosphorus (30). Responses to sulphur applications in this area have been obtained on various soil types and with various pasture species including subterranean clover (*Trifolium subterraneum* L.) (14), *Phalaris tuberosa* L., white clover (*Trifolium repens* L.) (3) and annual medics (*Medicago* spp.) (15). Hilder (14) found that applications of 30 to 60 kg per ha, applied as sulphates, were appropriate for the soils he examined. On these soils, the residual value of the phosphorus in superphosphate exceeded that of the sulphur component.

The southern areas of New South Wales appear to be less sulphur deficient than the Northern Tablelands, although some soils respond strongly to sulphur applications (21,31). Of twenty-one sites tested (31) eight responded to sulphur, five in the Monaro district, two in the Southern Tablelands and one in the Lachlan region.

In a study on the residual value of superphosphate at eighteen centres on the Southern Tablelands (23) only marginal responses to sulphur were obtained. In many instances yields of pasture were not lifted but the nitrogen contents were. Spencer *et al.*, (31) found the residual value of the phosphorus component of the superphosphate was greater than that of the sulphur component on most of the soils studied, especially on the red basaltics. It is interesting to note that sulphur shortage may limit organic matter accumulation of such soils (38).

Few reports have been made of sulphur responses in coastal areas due in part to sulphur accession in rainfall being higher near the sea (Table 4-4) and partly to the lack of active research in the area. Small increases in yield of subterranean clover have been obtained in Lower Hunter Valley soils where phosphorus deficiency was corrected first (35). In the Upper Hunter Valley, a large number of soils were found to be responsive to sulphur (24). Reports of sulphur deficiency in cereals in Australia have been rare, but sulphur responses in oats on a virgin basaltic soil have been obtained on the Central Tablelands (7).

Victoria

Recorded sulphur responses in Victoria have been mainly confined to the Gippsland area (8). In this region widespread deficiencies of both sulphur and phosphorus occur. On an old pasture with a long top-dressing history, response to gypsum alone was equivalent to that obtained from superphosphate, again demonstrating the variable residual effect of the phosphorus and sulphur components. In these trials, some responsive sites were located within a few kilometres of the sea, showing that, contrary to popular opinion, accession from this source was insufficient for maximum plant growth.

In a test on thirty-two virgin soils collected from southwestern Victoria to

TABLE 4-4
Sulphur accessions in rainfall in Australia and other selected areas of the world

Location	S addition (kg per ha per yr)	Reference
Western Australia	1.4 to 6.7	(10)
Victoria (coastal)	7.8	(17)
(inland)	<2	
N.S.W. Northern Tablelands	<1	(30)
Tasmania (coastal)	13.9	Martinick (pers. comm.)
(inland)	3.0	
Scandinavia	3	(11)
Nebraska, U.S.A.	4.3	(6)

Southern Queensland (22), twenty-four were deficient in sulphur. Nine of the eleven Victorian soils were sulphur deficient, and all were phosphorus deficient. The soils were from southwestern, central and nothern Victoria but did not include any from the Gippsland region. They were virgin soils and it may well be expected that their responsiveness would be much reduced following periods of regular cultivation and superphosphate application.

Recent work (Peverill, Briner and Walbran, pers. comm.) suggests that soils of low sulphur status occur near Melton and Bonny Doon, while others of moderate to low status occur near the South Australian border at Edenhope. Publication of results of the National Soil Fertility Survey should clarify the situation with regard to sulphur responses. Of ninety-eight areas tested for sulphur in Victoria responses were recorded on only four sites.

Tasmania

Early work was done by the Tasmanian Department of Agriculture (Stackhouse, pers. comm.). More recently a comprehensive study of sulphur in Tasmania (Martinick, pers. comm.) has been carried out. In this study, field and pot experiments have shown very significant responses to sulphur on podzolic coversands in the northeast of Tasmania near Bridport. This soil extends over a wide area along the northeast and eastern coast of Tasmania. Pot experiments suggested that several other soils, including podzols on sandstone and mudstone and lateritic podzols, are potentially sulphur deficient. Many of the areas shown responsive in these experiments occur near the coast where up to 10 kg S/ha/yr falls in the rain. Accession falls off rapidly with movement inland, and responsive sites on krazonzems near Deloraine and lateritic podzols near Longford (Stackhouse, pers. comm.) receive rainfall inputs of from 3 to 5 kg S/ha/yr.

South Australia

Sulphur responses in South Australia, as in Western Australia, appear to be confined to sandy soils in higher rainfall areas. Pastures on shallow sandy solonetzic soils near Keith have been shown to be responsive but responses vary seasonally depending on rainfall and fertilizer practice. Nicolson (25) observed a delicate sulphur balance when fine gypsum was used as a source of sulphur. Responses were increased in years of good winter rains and adequate growth (Powrie, unpublished data). In years of above average rainfall, responses were reduced by severe leaching losses, while in years of below average rainfall, yields were more limited by moisture shortage than sulphur supply. The variability of rainfall in the upper southeast of South Australia (34) makes it hard to predict sulphur responses in this area.

Responses to sulphur have been obtained on a podzolic soil west of Penola (5). Pasture that had received from 1 to 1.5 tons of superphosphate prior to 1962, when applications ceased, responded to reapplication of fertilizer in 1964. Greater responses to sulphur than to phosphorus were obtained, again demonstrating variable residual values of the main components of superphosphate. Recent work (Clarke and Lewis, pers. comm.) confirm responses on sandy soils in the southeast of South Australia, but again variable responses occur from year to year with variations in rainfall. Apart from the podzolic soils near Penola, which extend into western Victoria, the solodized solonetz soils near Keith, and the lateritic podzols near Mt Compass, no other South Australian soils respond to sulphur. Of twenty-nine sites tested in wheat and pasture areas during the National Soil Fertility Survey, none responded to sulphur applications (French, pers. comm.).

Western Australia

Sulphur deficiency is widespread on the sandy soils occupying approximately two-thirds of the southwestern region of the state (3). Most of the rain falls in the cooler months of the year (May to September) when evaporation is low, and there is a strong tendency for the coarse-textured soils to be leached.

Coolup sands on the Swan Coastal Plain respond to sulphur (3). These soils retain sulphur poorly and leaching losses are severe (16). On them, the residual value of the phosphorus component of superphosphate is higher than that of the sulphur component (3), but significant phosphorus losses also occur (16). Many of the forty-two soils examined in a pot experiment (3) responded to sulphur but several did not. Soils subject to occasional flooding by the slightly saline Murray River, low lying soils of the Boyanup and Turkey Flat series (19), those irrigated by the Waroona irrigation scheme and an estuarine alluvium failed to respond.

Many of the soils tested were near the coast but measured sulphur accessions in Western Australia have been low (Table 4-4). The coarse-textured soils combined with intense winter rainfall provide the ideal situation for regular annual leaching losses to occur.

PREDICTING RESPONSIVE SOILS

Attempts to relate soil type, parent material, environmental characteristics and other similar factors to sulphur responsiveness have met with limited success. Many factors contribute to this. Plant available sulphur is mainly in the sulphate form and this is in a dynamic state, being weakly held by most soils. The main reserve of sulphur is held in organic matter and factors affecting rates of accumulation and mineralization markedly affect sulphur availability. The root distribution of different plants may also influence their response to sulphur on different soils.

An early attempt to relate sulphur responses to various measurable factors was carried out on thirty-two virgin soils collected from Eastern temperate Australia (22). No relationship was found between responses and climatic and soil factors apart from pH. In an examination of forty Northern Tableland soils from New South Wales (30) no relationship between pH and sulphur response was found. The pH range of these soils was narrower than those tested in the previous survey, and the pH relationship noted earlier may have been unduly influenced by a small number of soils of high pH (> 8.0).

The colour of basaltic sedentary soils in northern New South Wales was found to be significantly correlated with sulphur status ($r = 0.92$), reddest soils being most deficient (29). There was no relationship with soil pH or redox potential. The red soils were probably high in exposed iron oxide, reflecting better aeration and drainage, and greater sulphur leaching. Such information, however, appears to be of limited value in predicting sulphur responses.

Adsorbed sulphate can be an important source of sulphur to plants and is a mechanism for retention against leaching. Soils with a high adsorption capacity might be expected to be less responsive than those with a low adsorption capacity.

Sulphur responsiveness on Dorrigo soils (20) was least on soils from the highest rainfall areas which also had the greatest sulphur adsorption capacity and were able to retain sulphate against leaching. Rainfall and sulphate adsorption capacity have been related but the relationship varied between soil types (4). The sulphur adsorption capacity of a soil appears to be an important factor in determining its sulphur responsiveness.

The influence of adsorption capacity and rainfall can be complicated by other factors. Negative correlations between rainfall and pH, and rainfall and sulphur response and a positive correlation between rainfall and sulphate adsorption were found on Dorrigo soils (20). This may not indicate a cause and effect relationship between sulphur responsiveness and the measured parameter. Sulphate adsorption occurs by reaction with surface aluminium and perhaps iron (13) and for some soils (4) the amount of reactive aluminium has been closely related to the pH, and both of these to rainfall. Sulphur accessions in rainfall could also influence the relationships. Factors other than adsorption capacity can be important in determining the sulphur status of these soils.

Till *et al.* (33) have shown that of a total soil sulphur pool of 245 kg/ha, 100 kg is non cycling and only 10 kg/ha/year is in available form. This data emphasizes the importance of the organic S → sulphate S process in determining the available sulphur status of the soil and hence sulphur responsiveness. It seems likely that since the S:P ratio in superphosphate is 1.3:1 and in organic matter is 2:1 (38), sulphur supply may limit organic matter accumulation in pastures fertilized with superphosphate. This in turn may limit the pool of organic sulphur available for mineralization. The mineralization rate of sulphur from organic forms is slow relative to nitrogen under field conditions. Plant material with an average N:S ratio of 15:1 ends up in soil organic matter with an average ratio of 7.4:1. This slower release of sulphur relative to N, coupled with the fact that most (96 per cent) of the soil sulphur is in the non-sulphate form (33) points to a blockage in the availability of sulphur to plants.

Many chemical assessments of sulphur status in Australian soils have been tested (32,36,37), but the transient nature of soil sulphate levels makes them difficult to apply except for short-term predictions. A recent development in this field has been the use of charcoal in the extracting procedure (Briner *et al.*, pers. comm.). The technique has been applied to a large number of Victorian soils with promising results.

Whether or not a soil is responsive to sulphur may also depend on the type of crop grown. The depth of root distribution of lucerne, for example, is much greater than for subterranean clover yet these two species are frequently sown on similar soils. The deeper penetration of lucerne roots may lead to it tapping reserves of leached or adsorbed sulphate which would not be accessible to subterranean clover. Such an effect was observed by the late J. K. Powrie in unpublished data from experiments on sandy soils in South Australia. The limitations of moisture and nitrogen that frequently restrict cereal growth under field conditions in most cropping areas of Australia probably overshadow the sulphur effects. Sulphur responses are rarely measured under these conditions although soil S levels may be low. Again the root systems of cereal plants are large and uptake of nutrients can occur from greater depths than for annual legumes.

In view of the many factors affecting sulphur responses it is not surprising that tests for detecting responsive soils have generally failed. Of the many responses to sulphur in the field, few, if any, have occurred in areas of below 500 mm rainfall. Those that have (notably in South Australia) have occurred in years of above average rain and responses have been poor or negative in years of below average rain. All field responses noted by the authors from Eastern Australia have occurred in regions where rainfall has been considerably higher than 500 mm, with areas of the Northern and Southern Tablelands of New South Wales, Gippsland and Tasmania recording averages as high or higher than 750 mm (see Figure 4-4). The position is similar in the southwest of Western Australia. From the data currently available the authors have produced a map outlining areas of temperate Australia within which sulphur responses might reasonably be expected (Figure 4-4). It has been based on observed field responses and extrapolated to similar soil types within similar rainfall areas.

Soils of the drier agricultural areas of Australia frequently have high adsorption capacities as a consequence of their higher clay content, such soils develop under less severe leaching conditions. Lower leaching losses, higher adsorption and a limitation on plant growth imposed by moisture contribute to their apparent sulphur sufficiency.

Acknowledgements. The authors are indebted to the many agriculturalists throughout Australia, too numerous to mention individually, who gave information which allowed them to compile the sulphur response map presented in this paper.

REFERENCES

(1) Anderson, A. J. and Spencer, D. (1950). *Aust. J. Sci. Res. Ser. B.*, **3**: 431.
(2) Banath, C. L. (1969). *Aust. J. Agric. Res.*, **20**: 697.
(3) Barrow, N. J. (1967). *Sulphur Inst. J.*, **3**: 2.
(4) Barrow, N. J. (1970). *Proc. XI Int. Grassld Congr.*, p. 370.
(5) Burford, J. R. (1968). *Agricultural Research in the South East of South Australia and Adjacent Areas of Victoria*, CSIRO, Melbourne, p. 29.
(6) Coleman, R. (1966). *Soil Sci.*, **101**: 230.
(7) Crofts, F. C. and Blair, G. J. (1966). *J. Aust. Inst. Agric. Sci.*, **32**: 142.
(8) Drake, F. R. and Curnow, B. C. (1967). *J. Dep. Agric. Vict.*, **65**: 229.
(9) Drake, F. R. and Curnow, B. C. (1968). *Sulphur Inst. J.*, **4**: 10.
(10) Drover, D. P. (1960). *J. Roy. Soc. W.A.*, **43**: 81.
(11) Eriksson, E. (1960). *Tellus*, **12**: 63.
(12) Glynne, M. D. (1929). *Proc. R. Soc. Vict.*, **42**: 30.
(13) Harward, M. E. and Reisenauer, H. M. (1966). *Soil Sci.*, **101**: 326.
(14) Hilder, E. J. (1954). *Aust. J. Agric. Res.*, **5**: 39.
(15) Hilder, E. J. and Spencer, K. (1954). *J. Aust. Inst. Agric. Sci.*, **20**: 171.

(16) Hingston, F. J. (1959). *J. Aust. Inst. Agric. Sci.*, **25**: 209.
(17) Hutton, J. T. and Leslie, T. J. (1958). *Aust. J. Agric. Res.*, **9**: 492.
(18) Leeper, G. W. (1935). *Proc. R. Soc. Vict.*, **47**: 225.
(19) McArthur, W. M., Bettenay, E. and Hingston, F. J. (1959). CSIRO Aust. Div. Soils, Soils and Land Use Ser. No. 31.
(20) McArthur, W. M. and Spencer, K. (1970). *Aust. J. Exp. Agric. Anim. Husb.*, **10**: 196.
(21) McLachlan, K. D. (1952). *Aust. J. Agric. Res.*, **3**: 125.
(22) McLachlan, K. D. (1955). *Aust. J. Agric. Res.*, **6**: 673.
(23) McLachlan, K. D. and Norman, B. W. (1962). *Aust. J. Agric. Res.*, **13**: 836.
(24) Mitchell, A. L. (1956). *J. Soil Conserv. Serv. N.S.W.*, **12**: 140.
(25) Nicolson, A. J. (1970). PhD thesis. University of Adelaide.
(26) Powrie, J. K. (1963). *J. Dep. Agric. S. Aust.*, **66**: 294.
(27) Roseworthy Agricultural College, Annual Progress Report 1882.
(28) Rossiter, R. C. (1952). *Aust. J. Agric. Res.*, **3**: 7.
(29) Spencer, K. (1966). *Aust. J. Soil Res.*, **4**: 115.
(30) Spencer, K. and Barrow, N. J. (1963). CSIRO Aust. Div. Plant Industry Tech. Pap. No. 19.
(31) Spencer, K., Bouma, D., Moye, D. V. and Dowling, E. J. (1969). *Aust. J. Exp. Agric. Anim. Husb.*, **9**: 310.
(32) Spencer, K. and Freney, J. R. (1960). *Aust. J. Agric. Res.*, **11**: 948.
(33) Till, A. R., May, P. F. and McDonald, I. W. (1970). Muth, O. H. (ed.), *Sulphur in Nutrition*, Avi Publishing Co., U.S.A.
(34) Trumble, H. C. (1948). *J. Dep. Agric. S. Aust.*, **52**: 155.
(35) Weatherall, R. S. (1963). *Agric. Gaz. N.S.W.*, **74**: 83.
(36) Williams, C. H. and Steinbergs, A. (1962). *Pl. Soil*, **17**: 279.
(37) Williams, C. H. and Steinbergs, A. (1964). *Pl. Soil*, **21**: 50.
(38) Williams, C. H. and Donald, C. M. (1957). *Aust. J. Agric. Res.*, **8**: 179.

The Occurrence of Sulphur Deficiency in New Zealand

T. W. WALKER and P. E. H. GREGG

Although sulphur (S) responses were obtained in the 1920s, the extent of the problem was not appreciated until the 1950s when large areas of unimproved grasslands in the South Island were shown to need S in order to stimulate nitrogen fixation by introduced clovers. This early work has been adequately reviewed recently (15,17).

Stimulated by Lobb's pioneering work (12) Walker *et al.* (26) laid down field experiments in Canterbury, as a result of which the following hypotheses were developed:

> Most of the S in well-drained soils is in organic forms with relatively constant N:S ratios. Grasses are able to take up about 90-95% of the N and S mineralised from soil organic matter leaving clovers dependent mainly on outside sources of S. The amount of S needed by clover depends on climatic and other factors governing the amount of N fixed, as 1 kg S is needed to fix 10-15 kg N. Atmospheric returns of S are relatively low over most of New Zealand and very low in inland areas. Responses to S of introduced clovers should therefore be ubiquitous away from the coast. Because much of the applied S accumulates in soil organic matter and only slowly mineralises, fertiliser-S should be necessary during the phase of organic matter accumulation under permanent pastures, and only after reaching an apparent steady state should the requirements decline. Soils ploughed for cropping should be able to meet the S requirements for crops if they can meet the N requirements, as mineralisation of organic N and S should proceed at similar rates.

There are now more soil data and field experiments which substantiate, refute or lead to modification of the above statements and these will be examined by considering the occurrence of S deficiency on: 1. undeveloped and 2. developed pastures.

UNDEVELOPED PASTURES

Weathering from parent rocks

As legumes responded markedly to S even on skeletal soils in inland Canterbury containing much gravel and stone, it was concluded that primary sulphides are quickly oxidized to sulphate and leached where drainage is good. Tillman (pers. comm.) has provided confirmation from analyses of weathering rocks sampled from quarry faces and exposed cliffs. Table 4-5 presents data from a quarry face where P fractions and mineralogy of the same samples were also studied (21).

TABLE 4-5
Total S, sulphate-S, sulphide-S, in a greywacke soil-rock weathering sequence in North Auckland

Depth	Total S* (ppm)	Sulphate-S** (ppm)	Sulphide-S† (ppm)
0–10 cm	432	47	< 5
30–45 cm	275	73	< 5
3 m	273	224	32
9 m	123	97	33
14 m	69	27	48
18 m	718	64	640

* Total S determined by dry ashing with $NaHCO_3 + Ag_2O$.
** Sulphate-S determined by extraction with phosphate solution (500 ppm).
† Sulphide-S determined by decomposition with $HNO_3 + Br_2$.

In the unweathered rock at 18 m, sulphide-S was the dominant component. The weakly weathered rock at 14 m had lost most of its S but a small amount of sulphide-S persisted. At 9 m and 3 m total S increased in the form of SO_4-S, probably accumulating by downward leaching from the top-soil where most of the S is organic. At 3 m and 9 m the feldspars (dominant at 14 and 18 m) had weathered mainly to amorphous material and kaolinite. Rock profiles in the South Island show similar dramatic losses of sulphide-S at considerable depths.

Tillman also perfused rock ground to less than 60 mesh and stones less than 1 cm in a perfusion apparatus. About 1 per cent of the total S (750 ppm) was oxidized in a week from 20 g of the fine material when perfused with water, about 30 times faster than the stones. Rocks containing sulphide, decomposable by concentrated HCl oxidized twice as fast as rocks containing pyrite-type compounds insoluble in HCl. Inoculation with S-oxidizing bacteria trebled the rate of oxidation. These data support the view that primary sulphides in unweathered rocks, would not persist for very long even under conditions of weak weathering and leaching.

Additional field evidence was provided from a study of a chronosequence of soils on sand-dunes (20) where total S declined rapidly in the early years and then increased in a parallel fashion with C and N, presumably due to atmospheric accessions and subsequent accumulation in organic matter.

Inorganic sulphate

A detailed review of the S status of unimproved soils (17) has shown the S responsive soils in New Zealand to be either low in clay in the upper horizons (e.g., gley podzols and yellow-brown pumice soils) or to have dominant 2:1 lattice type clays giving low anion retention (e.g., brown-grey earths, yellow-grey earths, high country yellow-brown earths, recent soils from greywacke loess and alluvium and related soils). Such soils almost invariably have low contents of adsorbed sulphate (phosphate exchangeable) throughout the profile. Most of the soils derived from volcanic rocks (except weakly weathered yellow-brown pumice soils very low in clay) tend to accumulate much adsorbed sulphate (> 40 ppm) particularly below the top 20 cm. The dominant clays in the soils are allophane, 1:1 lattice (kaolinite, halloysite) and gibbsite. Metson (17) considers that the massive concentrations of adsorbed sulphate in some deep horizons are unlikely to contribute significantly to the growth of shallow rooting species in permanent pastures as these high values are also usually associated with high acidity, low base saturation and a high content of aluminium-rich clay—all of which discourage deep root development.

Where an acute deficiency of P or Mo or heavy metal toxicity related to low pH may be restricting plant growth, adsorbed sulphate may accumulate in spite of low atmospheric returns and this may meet the S requirements for establishing clovers when developing such soils.

Thus a major reason for lack of or slight responses of introduced legumes to S is likely to be the presence of appreciable water-soluble or adsorbed sulphate.

Atmospheric returns of sulphur

A. H. Horn (pers. comm.) has measured the amount of SO_4-S collected in precipitation in two transects across the South Island. Results for one of these transects from Lincoln (east) to Kumara (west) are given in Table 4-6 (mean values for 3 to 4 years).

A similar transect across Otago gave values of 4 kg/ha/yr at the coast, declining to 0.6 kg in Central Otago and rising again to 9.7 kg at Haast (1.5 km from the coast). Sodium and chlorine follow similar patterns denoting the probable maritime origin of much of the sulphur. There are no similar transects for the North Island, but Miller (18) measured inputs (including dust) of 10 kg S/ha/yr near Wellington. Values of near 15 kg are regularly reported near Auckland (19).

Although evaluation of direct plant and soil absorption of SO_2 has not been reported in New Zealand, these quantities are likely to be related to returns

TABLE 4-6
Sulphate-S in precipitation (excluding dust)

Site	Km from coast		SO_4-S (kg/ha/yr)
	From east	From west	
Lincoln	22		4.0
Darfield	42		1.7
Torlesse	67		1.9
Cass	100		1.1
Bealey	109		1.0
Arthurs Pass	122	51	3.4
Taipo		29	3.7
Kumara		3	5.6

in precipitation and therefore to be considerably less inland than on the coast. Even if these values equalled those in precipitation, many inland areas would be receiving annually only about 2 kg S/ha from outside the ecosystem, and as some of these additions would occur in the winter during periods of low pasture growth, leaching losses could easily occur on soils with little ability to retain adsorbed sulphate.

Indirect evidence of the low contribution of direct absorption of SO_2 by plants and soil from the air was afforded in a mowing trial examining residual effects of gypsum over a four-year period, on a soil with little or no adsorbed sulphate in an area where return in precipitation is about 1 kg S/ha/annum (26). The pasture responded significantly to as little as 5.8 kg S/ha applied as gypsum.

Of the 5.8 kg S applied initially 77 per cent was recovered over four years. During this period the control plot yielded 6.6 kg S/ha or 1.6 kg/ha/yr. Even if the whole of this had come from the atmosphere and the recovery by the pasture of atmospheric returns (all kinds) had been 80 per cent, these returns would not have exceeded 2 kg/ha/yr, suggesting that direct absorption of SO_2 by plants and soil was less than returns in precipitation. Such a calculation ignores microbial effects but confirms the low values from all atmospheric sources in inland areas of the South Island, likely to be a most important factor limiting biological N-fixation over much of New Zealand.

Subsequent field trials have generally confirmed the original hypothesis which predicted small responses if any, of introduced clovers to fertilizer-S near the coast and widespread responses inland. Adams and Walker (unpublished) in the mid-fifties found no responses to S in several trials on Banks Peninsula (virtually surrounded by the sea and rainfall varying from 500 to 1,500 mm). The conclusion reached was that atmospheric returns (5-10 kg S/ha/yr) were adequate for plant requirements, but some of the soils derived from basalt may have contained appreciable adsorbed sulphate.

In association with the rain gauges collecting precipitation on the Otago transect, eleven trials measuring S responses on previously untreated soils indicated no responses of clovers on coastal trials. Responses developed 6 km from the coast (Cossens, pers. comm.). More recently in a series of fourteen field trials around the Otago Peninsula, Feyter (pers. comm.) obtained responses to S at only two sites, both recording the lowest rainfall in the series. Risk (pers. comm.) in a series of six trials in Western Southland, obtained good sulphur responses to the west of the Takitimu mountains (wetter side) and nil to very slight responses on the eastern coastal side.

Cuff (6), on steepland soils in South Canterbury, obtained no responses to S in two trials 6 km from the coast, which he did not attribute entirely to returns in precipitation (2-3 kg/ha/yr) but suggested that sea fog, a common occurrence in this area, may contribute to the external supply of S. This stony steepland soil contained appreciable amounts of adsorbed sulphate in the top 5 cm and had high sulphur retention characteristics.

However, some excellent S responses have been recorded in coastal areas. Douglas (pers. comm.) in Oamaru and McLeod (pers. comm.) at Timaru, have both reported responses. More recently Gregg (unpublished) in a sequence of four loess-derived soils on Banks Peninsula, obtained a significant response to S (1 per cent level) at a site only 400 m from the coast. On the West Coast, S responses were reported when introducing clovers into 'pakihi' soils (16) and O'Connor (pers. comm.) has recorded S responses in recent soils.

As most of the above trials were assessed visually with few supporting soil and herbage data, it is difficult to explain some of these results. Pasture productivity at trial sites is of major importance. Where clover growth is poor due to any factor other than S deficiency (e.g., climate) requirements for S will be low and where atmospheric returns are border-line, responses to fertilizer-S might be obtained in a year of high productivity but not in a year of low productivity.

In inland areas of the South Island, about 150 field trials conducted since the 1950s, have confirmed the widespread nature of S responses. Again there have been exceptions. There is a small area of soils in the semi-arid climate of Central Otago (300-500 mm) where carbonate and water-soluble sulphate accumulate in lower horizons. Such soils will grow unirrigated lucerne without responding to S (3).

Recently, in thirty trials on steepland soils in Canterbury, Gregg (unpublished) failed to obtain responses to S at two sites, both of which responded well to P and Mo. Similarly Cossens (pers. comm.) obtained a response of clovers to P but not to S on Teviot soils of the inland Otago plateau. The soils at these three sites, previously untopdressed tussock country and well away from the coast, were of relatively low pH (5-5.3), and had medium and high S retention characteristics in the topsoil and subsoil respectively. In the Canterbury sites, adsorbed SO_4-S levels (20-30 ppm) could explain the lack of S responses. At the Teviot site and one Canterbury site responses appeared in the second year.

Factors of soil acidity may also limit S responses. Walker *et al.* (24) reported a possible manganese toxicity on clover and in a field trial (25) clovers were actually suppressed by the application of gypsum, which increased Mn uptake on an acid (pH 5.7), poorly drained soil high in available manganese. Some of the rare cases where no response is obtained to S in unimproved tussock country in the South Island may be due to this cause.

For many years there was a puzzling lack of responses to S reported from the North Island even in areas well away from the coast. This may be due partly to the fact that most of the trial work was concentrated on developed pastures adequately supplied with S in top-dressings of superphosphate. More recently, widespread occurrences of S deficiency have been reported on undeveloped country, e.g., on the yellow-brown pumice soils (23), in Northland, Gould and Ludecke (pers. comm.), and on soils in the Wanganui and Rangitikei districts (2). No trials have been reported from unimproved pastures on soils likely to have naturally high levels of adsorbed-SO_4, which exist over large areas of the North Island.

Mineralization of organic S

The only study on mineralization of organic S since the early work (28) is that of Tan's (22), which followed the mineralization of N and S in a pluvosequence of weakly weathered zonal soils derived from mica-schist. In a series of field trials on these same soils Ludecke (14) studied the effects of S, P and Mo on clover establishment and growth. Responses to S were found on all but the driest member of the sequence. The rainfall and some properties of the 0-10 cm layer of the soils are given in Table 4-7. This shows the declining pH and most probable number of nitrifiers, increasing percentage C and relatively constant total N:S ratio (mean = 8.7) as precipitation increases. Tan incubated the soils (60 per cent W.H.C.) at 25°C measuring levels of NH_4-N, NO_3-N and SO_4-S (phosphate extractable) over a period of 80 days. The forms of mineral-N produced (but not the amounts) were governed by the numbers of nitrifiers which in turn depended on soil pH. Mineral N:SO_4-S ratios (Table 4-8) were very low at zero time but quickly increased

TABLE 4-7
Soils, precipitation and some properties (0 to 10 cm) of a pluvo-sequence

Soil	Zonal classification	Precipitation (mm)	pH	C (%)	N (%)	S (%)	Nitrifiers M.P.N./g
Conroy	B.G.E.	325–350	6.4	1.56	0.16	0.019	1,700
Omakau	B.G.E.	300–400	6.1	2.86	0.26	0.032	1,000
Pigburn	Transitional B.G.E.-Y.G.E.	400–450	6.3	2.34	0.22	0.025	1,500
Arrow	Y.G.E. (steepland)	600	5.8	2.99	0.27	0.030	900
Blackstone	Y.G.E.	700	5.7	2.73	0.26	0.031	600
Wehenga	Transitional Y.G.E.-Y.B.E.	900	5.5	3.12	0.32	0.036	120
Carrick	High country Y.B.E.	1,000+	5.1	3.25	0.27	0.027	50

TABLE 4-8
Ratios of mineral N:SO_4-S at 0 and 80 days incubation and Δmineral N:ΔSO_4-S after 80 days

Soil	Mineral N:SO_4-S		Δ mineral-N / Δ SO_4-S
	0 days	80 days	
Conroy	1.5	6.2	19
Omakau	1.8	9.0	24
Pigburn	1.8	12.7	39
Arrow	1.8	13.5	33
Blackstone	1.5	13.3	42
Wehenga	1.5	11.8	25
Carrick	1.7	7.4	61
Mean	1.6	10.6	35

within the first 20 days to values exceeding 10 for all but the Conroy, Omakau and Carrick soils. As shown by several Australian workers (1,8,29) organic N mineralized much faster than organic S as judged by △mineral N: △SO_4-S ratios. Tan showed that the mineral N:SO_4-S ratio of living roots was 1.1 compared with 1.4 in the surrounding soil. Roots may influence net mineralization of N and S from native soil humus (9). The significant values for plant growth are likely to be the actual mineral N:SO_4-S ratios after 80 days. The mean value (10.6) is within the range (8-12) commonly found for N:S uptake ratios in pastures thus strengthening the original hypothesis on rates of mineralization. The Conroy soil with the lowest ratio of 6.2 is the soil most likely to provide sulphate in excess of that required by grasses using all the mineral-N. It is the one soil on which lucerne does not respond to fertilizer-S as it contains SO_4-S in the subsoil (3).

Tan also examined the effect of liming on mineralization of N and S. Liming caused very marked increases in SO_4-S for the first four members of the sequence and much smaller increases for the three most acid soils. Liming increased levels of mineral-N in all soils but differences between soils were small. Thus liming gave low mineral-N:SO_4-S ratios in the four least acid soils and considerably higher ratios in the three most acid soils. This could explain the small responses to S on some soils after liming.

Developed pastures

Ostensibly used to correct P deficiency, superphosphate also corrects S deficiency. Where applications of about 300 kg or more of superphosphate/ha/yr are needed, requirements for S should automatically be satisfied in most conditions. On P rich soils, natural or fertilized, sulphurized-superphosphate is recommended. The need for sulphur should continue on improved pastures because of its accumulation in organic matter, losses by leaching, evolution of H_2S, and removal in animal products.

Accumulation of organic S

Early evidence for the accumulation of S in organic matter under pastures treated with superphosphate (27) showed that total S in the top 20 cm increased from 290 kg/ha in the virgin soil to 670 kg/ha after twenty-five years; C:N ratios fell slowly from 33 to 11 whereas C:S ratios fell more rapidly from 280 to 90.

Gregg (unpublished) using gypsum labelled with S^{35}, found that incorporation of fertilizer-S into organic S, probably by micro-organisms, was very rapid. In a field trial on a three-year-old pasture, approximately 40 per cent of the fertilizer-S in the 0-5 cm horizon had been converted to organic S only six weeks after application.

On several North Island soils growing good quality pasture, accumulation of organic N, S and P was general in the top 7.5 cm, but in soils which had reached an apparent steady state at the time of ploughing, net mineralization occurred between 7.5-15 cm (11). In eight of the ten series, immobilization exceeded mineralization.

Most recently, Burke (5) reported a nearly two-fold increase in organic S on irrigated pastures in Canterbury top-dressed with superphosphate for twenty-five years.

Once an apparent steady state for organic matter has been reached, no more S will be 'fixed' in organic forms and presumably only as much S will be required as is lost from the system. Wherever atmospheric returns approach or exceed these losses, requirements for fertilizer-S will be minimal.

Leaching losses

There have been few, if any, direct measurements of losses of sulphate by leaching in the field. Sulphate leaches easily from yellow-brown pumice soils in laboratory experiments (10), and in field experiments (23) sulphur responses occurred in the spring after autumn applications of 250-375 kg super/ha, presumably due to leaching of sulphate during the winter. On an undeveloped gleyed podzol (2,000 mm) marked lateral movement of S^{35} labelled gypsum has occurred due to impeded drainage (7). However, losses have been small from lysimeters growing herbage at Otara (19).

Gregg (unpublished) studied the downward movement of sulphate from application of S^{35} labelled gypsum at nine sites in the field. At two sites which received 250 mm of rain in the six weeks after application, sulphate had penetrated to 30 cm in a silt loam and 60 cm in a sandy loam. It is difficult to evaluate effects of downward leaching without knowledge of depth of rooting and uptake. Gregg (unpublished) inserted S^{35} labelled gypsum in gelatin capsules at 7.5, 22.5, 37 and 52 cm depth. Uptake by grasses and clovers occurred from all depths in both the sandy loam and silt loam with greater recovery at lower depths in the sandy loam. Subsequently, uptake by grasses and clover from 100 cm depth in the sandy loam, but only 52 cm in the silt loam (Table 4-9), was reported (4).

TABLE 4-9

Uptake by grass and clover of S^{35} labelled sulphate placed at different depths in two soils of different texture. Autumn cut

Depth of placement of labelled sulphate cm	Sandy loam		Silt loam	
	Grass	Clover	Grass	Clover
	(*dpm × 10^4/g herbage)			
22	617	153	540	230
52	530	430	21	40
75	113	33	0	0
100	324	20	0	0

* Disintegrations per minute.

Leaching losses will clearly be affected by sulphate retention capacity, which has been shown to vary widely in New Zealand soils (Saunders, pers. comm.), and the use of phosphate fertilizers and lime.

It is difficult to test, by field trials, the hypothesis that S requirements will continue but eventually decline on improved pastures, without corresponding detailed soil studies, which are generally lacking.

Cossens (pers. comm.) found that only 20 kg S/ha/yr were required to maintain production of 11,000 kg dry matter/ha/yr on an irrigated pasture in Central Otago, and only 10 kg S/ha to produce 8,000 kg DM/ha on an irrigated brown-grey earth site. In a similar trial at Nelson, O'Connor (pers. comm.) found the requirements in the later years to be 20 kg S/ha/yr to produce 12,000 kg DM/ha. In lucerne cut for hay at Lincoln no responses to S were found over a three-year period on a shallow stony Eyre soil yielding 4,000 to 5,000 kg DM/ha/yr. Investigating the reason for lack of S responses over a four-year period on lucerne yielding 15,000 kg DM/ha on a weakly saline soil (Motukarara silt loam), Gregg (unpublished) found 20 ppm water-soluble SO_4-S to 40 cm depth rising to over 400 ppm at 75 cm.

On withholding superphosphate from pastures for one year in a large series of trials throughout New Zealand, Saunders (pers. comm.) reported no responses to S, indicating that the previous year's top-dressing was more than adequate. Sulphur responses are occurring after withholding superphosphate for two years on highly sulphate-retentive yellow-brown loams of Waikato under 1,200 mm rainfall. In earlier trials (13), after withholding superphosphate for six years, S deficiency was more severe than P deficiency.

REFERENCES

(1) Barrow, N. J. (1961). *Aust. J. Agric. Res.*, **12**: 644.
(2) Blackmore, L. W., Ludecke, T. E. and Hogg, D. E. (1969). *N.Z. Grassld Assoc. Proc.*, **31**: 87.
(3) Blakemore, L. C., Ludecke, T. E., Leamy, M. L. and Metson, A. J. (1969). *N.Z. J. Agric. Res.*, **12**: 333.
(4) Brash, D. (1973). BAgrSci (hons) dissertation, Lincoln College.
(5) Burke, A. S. (1973). *N.Z. J. Experi. Agric.*, **2**: 131.
(6) Cuff, J. R. I. (1973). MAgrSc thesis, Lincoln College.
(7) During, C. and Martin, D. J. (1968). *N.Z. J. Agric. Res.*, **11**: 665.
(8) Freney, J. R. and Spencer, K. (1960). *Aust. J. Agric. Res.*, **11**: 339.
(9) Freney, J. R. and Stevenson, J. K. (1966). *Soil Sci.*, **101**: 307.
(10) Hogg, D. E. and Toxopeus, M. R. J. (1966). *N.Z. J. Agric. Res.*, **9**: 93.
(11) Jackman, R. H. (1964). *N.Z. J. Agric. Res.*, **7**: 445.
(12) Lobb, W. R. (1954). *N.Z. J. Agric.*, **89**: 434.
(13) Lobb, W. R. (1962). *N.Z. Grassld Assoc. Proc.*, **20**: 50.
(14) Ludecke, T. E. (1966). *Proc. X Int. Grassld Congr.*, p. 856.
(15) Ludecke, T. E. (1969). *Proc. XII Tech. Conf. N.Z. Fert. Man. Ass.*, p. 85.
(16) McNaught, K. J. and During, C. (1970). *N.Z. J. Agric. Res.*, **13**: 567.
(17) Metson, A. J. (1969). *Proc. XII Tech. Conf. N.Z. Fert. Man. Ass.*, p. 73.
(18) Miller, R. B. (1968). *Trans. IX Int. Cong. Soil Sci.*, **4**: 323.
(19) Muller, F. B. and Shawcross, E. E. (1971-2). Annual Report N.Z. Fert. Man. Ass., p. 42.
(20) Syers, K. J., Adams, J. A. and Walker, T. W. (1970). *J. Soil Sci.*, **21**: 146.
(21) Syers, K. J., Williams, J. D. H. and Walker, T. W. (1970). *Soil Sci.*, **110**: 100.
(22) Tan, K. H. (1967). MAgrSc thesis, Lincoln College.
(23) Toxopeus, M. R. J. (1970). *Proc. XI Int. Grassld Congr.*, p. 345.
(24) Walker, T. W., Adams, A. F. R. and Orchiston, H. D. (1955). *Pl. Soil*, **6**: 201.
(25) Walker, T. W., Adams, A. F. R. and Orchiston, H. D. (1955). *N.Z. J. Sci. Tech.*, **36A**: 470.
(26) Walker, T. W. and Adams, A. F. R. (1958). *Pl. Soil*, **10**: 176.
(27) Walker, T. W., Thapa, B. K. and Adams, A. F. R. (1959). *Soil Sci.*, **87**: 135.
(28) White, J. G. H. (1959). *N.Z. J. Agric. Res.*, **2**: 255.
(29) Williams, C. H. (1967). *Pl. Soil*, **24**: 205.

The Occurrence of Sulphur Deficiency in Forests

F. R. HUMPHREYS, MARCIA J. LAMBERT and J. KELLY

Sulphur is of the same order of importance to trees as it is to most other plants. The level of sulphur found in tree foliage is similar to that of phosphorus. However, for the species examined in Australia (*Pinus radiata* (radiata pine), *P. elliottii* (slash pine), *P. taeda* (loblolly pine), *P. caribaea* (caribbean pine) and *Eucalyptus pilularis* (blackbutt)) total P and total S are not significantly correlated, except where superphosphate has been applied.

In *P. radiata* foliage, organic sulphur* was found to be highly correlated with total nitrogen (5) ($r = 0.960$). The ratio found ($S:N = 0.03$ on a gram atom basis) appears to be the same for *P. elliottii* and *P. taeda* whereas *P. caribaea* and *E. pilularis* seem to have a slightly different ratio (6). In view of the known biochemical association of these two elements in plants, this correlation reflects a true functional relationship. In the field situation, total foliar nitrogen has always been found to be equivalent to organic foliar nitrogen. However, when the sulphur supply is adequate, the pine accumulates, as sulphate, any excess of that sulphur required to balance the nitrogen available. Hence the sulphate-sulphur content of the foliage gives an estimate of the sulphur available.

In contrast with phosphorus and nitrogen, there do not appear to be any unequivocal demonstrations of growth responses to sulphur by trees in the forest situation, although they have been demonstrated in culture solution and pot experiments for many commercial species. Beaton (2) reported a sulphur response in the greenhouse on pumice soils for *P. ponderosa* and found some evidence of a response in the field. Ingestad (4) showed that there was a clear relationship between growth and sulphur for *P. sylvestris* in culture solution. In glasshouse pot studies with lateritic red earths and other Northern Territory soils, D. Cameron (pers. comm.) has shown a growth response to sulphur with *Khaya senegalensis* (African mahogany) and *P. caribaea.*

Lyle (7), also working with culture solutions showed that in solutions lacking in sulphur, *P. taeda* developed yellowing which started with the terminal needles and spread over the seedlings. Attempts to produce sulphur deficiency symptoms in *P. radiata* in sand culture in a glasshouse in the Sydney city area failed, presumably because of the SO_2 in the air (8), although a number of eucalypts readily developed symptoms in the same location

* Organic sulphur is the difference between total sulphur and inorganic or sulphate sulphur.

using the same cultural conditions and techniques (9). However, sulphur deficiency symptoms and some growth reduction were achieved by Afzal (1) working in Canberra using a vermiculite-perlite culture solution system. The symptoms he reported (needles light yellow in colour, length of needles decreasing towards the apex, terminal foliage chlorotic, the apical buds of all plants healthy but generally resinous) correspond with the symptoms noted in field examinations of some groups of trees found to have particularly low levels of sulphur in their foliage (S. W. Gentle, pers. comm.). Of particular importance in this connection is the production of resin from the apical bud, indicating the existence of lesions. Under some circumstances, these will allow the entry of fungal spores which are known to produce a die-back disease which is very common in these areas (S. W. Gentle, pers. comm.). Afzal (1) found a significant but not very great effect on height growth; however, diameter growth was severely curtailed.

The poor correspondence between glasshouse and field experience is not uncommon in forestry nutritional studies and the cause can probably be found in the very different associated environmental factors. The requirement for sulphur even under conditions of the highest attainable growth rates is only in the vicinity of 1 kg/ha/year, although at times between three to seven years of age, considerably more may be required. This is quite low compared with its concomitant nutrient, nitrogen, and is much more readily obtained by a pine forest from atmospheric sources that nitrogen appears to be. The foliage of *P. radiata* readily absorbs SO_2 from the atmosphere and the arboreal habit of trees is a considerable aid to this process. Some sulphur is supplied by rain even on the approaches to the western slopes of the Eastern Australian Tablelands (3). Sulphur is quite readily retained by the litter and is available to the trees from this source as a well-controlled, recycled nutrient.

Despite this, in many places in Eastern Australia, particularly on some parts of the tablelands where growth conditions otherwise appear to be ideal for *P. radiata*, there are patches of trees showing symptoms identical with

TABLE 4-10
Typical foliage analyses of *P. radiata* grown on granite or basalt soils

Location	Soil parent material	N %	P %	Al %	Ca %	Mg %	K %	Total S %
Green Hills	Granite	1.58	0.199	0.122	0.372	0.219	0.485	0.108
Nalbaugh	Granite	1.66	0.306	0.054	0.192	0.209	0.860	0.120
Hoad's Farm	Basalt	2.74	0.364	0.079	0.149	0.711	2.020	0.184

	Na ppm	Fe ppm	Mn ppm	B ppm	Zn ppm	Cl ppm	SO_4 S ppm
Green Hills	60	140	380	14	72	2	16
Nalbaugh	45	135	115	43	85	50	10
Hoad's Farm	70	225	240	38	40	650	0

TABLE 4-11
Sulphate-S levels in *P. radiata* foliage from various locations

Location	Mean Sulphate-S ppm S	% samples		S sorption maximum of soil (b) (μg/g)
		< 10 ppm Sulphate-S	< 25 ppm Sulphate-S	
Mt Gambier	518	0	0	0.37
Lidsdale S.F.	355	0	0	0.28
Newnes S.F.	290	0	4	32.03
Sunny Corner S.F.	241	0	0	10.09
Carabost S.F.	108	11	21	2.61
Nalbaugh S.F.	80	21	25	73.39
Green Hills S.F.	78	19	50	2.09
Mullion Range S.F.	64	27	50	0.38
Bondi S.F.	56	17	41	0.28
Mannus S.F.	21	80	80	4.86

those described by Afzal (1). These symptoms are found most frequently in fast growing plantations on soils derived from basalt and granite. Typical foliar analyses for *P. radiata* foliage from some of these soils are shown in Table 4-10. In vigorously growing trees these foliar nutrient levels indicate excellent mineral nutrition except for negligible sulphate-S. A site index/soil analysis/foliage analysis survey currently being carried out indicates that there are significant and marked differences between the foliage sulphate-S levels at various locations (Table 4-11). The location of the Mt Gambier forests ensures a strong maritime influence. Lidsdale, Sunny Corner and Newnes are each within a radius of 15 km of the Wallerawang Power Station. In addition to the SO_2 output from this station, fly ash containing Na_2SO_4, can from time to time be seen deposited on the foliage at Lidsdale and, less frequently, at Newnes. The remainder are watered by rainfall which generally comes from the southwest and has a very low sulphur content. The air contains only the basic SO_2 level with no industrial effluent. As may be expected, the frequency of occurrence of sulphate-S levels less than 10 ppm increases as the mean sulphate-S level decreases, reaching the very high figure of 80 per cent at Mannus State Forest, a site where dead-topping has seriously hampered the utilization of the crop. The sulphate adsorption characteristics of the soils (Figure 4-5) as determined by the Langmuir isotherm calculation show no relation with the foliage sulphate levels (Table 4-11).

Sulphate levels appear to be under strong genetic control. Foliage samples from three groups of *P. radiata* trees grown as cuttings from different clones were sampled on the same day and examined for sulphate-S content. A highly significant difference was obtained between the means of the individual

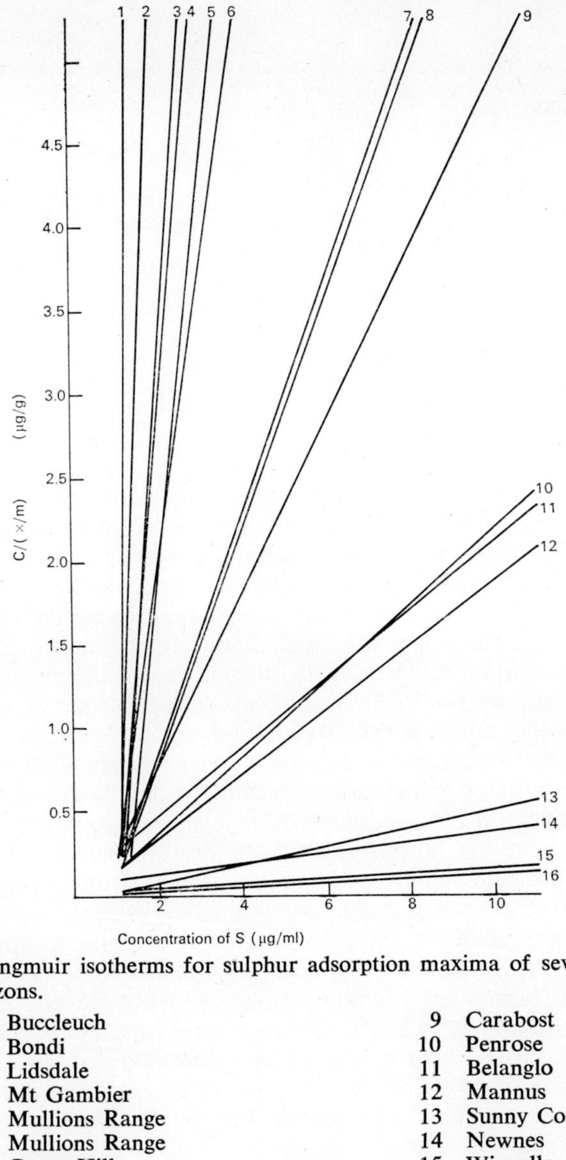

Figure 4-5 Langmuir isotherms for sulphur adsorption maxima of several soils, all 0-8 cm horizons.

1	Buccleuch	9	Carabost
2	Bondi	10	Penrose
3	Lidsdale	11	Belanglo
4	Mt Gambier	12	Mannus
5	Mullions Range	13	Sunny Corner
6	Mullions Range	14	Newnes
7	Green Hills	15	Wingello
8	Mellong Range	16	Nalbaugh

clones (mean sulphate-S: 296, 112, 46 ppm S). Also when examining the problems associated with a seed orchard which consisted of clonal material grafted onto randomly selected root stock, clonal differences in sulphate-S were found (Table 4-12).

In forestry we are probably looking at sulphur deficiency as a factor associated with the development of physiological malfunction. It is possible that sulphur deficiency leads to the development of a disease which affects

TABLE 4-12
Sulphate-S levels in *P. radiata* foliage of various clones of grafted material

Clone No.	Mean sulphate-S ppm S	% samples sulphate-S <10 ppm S
56	145	0
62	140	0
64	58	27
70	59	50
80	24	64
83	128	0
87	141	0
LSD (0.01)	66	

yield, rather than directly resulting in a yield reduction. One hypothesis concerning this disease is that various nutritional factors such as sulphur and/or boron deficiency, drought, and fissures caused by such physical agencies as hail and cockatoos can lead to the trees becoming predisposed to the entry of a fungus, *Diplodia pinea*, which then causes the subsequent symptoms. The disease does not occur every year and seems to affect trees mainly during the period of maximum height growth, that is, up to about age fifteen. It is difficult, therefore, to produce unequivocal results from field experiments designed to test the connection between sulphur deficiency and the disease.

In Table 4-13, the decline in foliage levels of sulphate-S with time from contrasting sites treated with various fertilizers can be seen. In the poorly sulphur-adsorptive soils (at Mullions Range and especially Buccleuch), the decline in foliage levels of sulphate-S with time is particularly marked. In contrast the foliar sulphate-S levels in the foliage from Wingello are constantly high. Table 4-14 indicates a corresponding decline in foliage nitrogen content with the decline in sulphate-S. This is a possible illustration of a limiting sulphur supply affecting the nitrogen uptake. The related mensuration data (bearing in mind that Mullions Range is a phosphorus deficient area) suggests the possibility of a height response (and increased diameter growth) when sulphur is not limiting in the foliage. The addition of gypsum to this poorly sulphur-adsorptive soil fertilized with superphosphate can be seen to maintain higher sulphur levels in the foliage than superphosphate alone. The results in Table 4-15 for a rock phosphate + S fertilizer instead of rock phosphate alone or superphosphate alone make this form of fertilizing look attractive. The difficulty in producing unequivocal results from field experiments designed to test any correlation between sulphur deficiency and disease is further exacerbated by the obvious necessity to continually apply sulphates to certain soils—a process of nil interest to forestry management.

There is considerable pressure in forestry circles to boost the production of timber with the use of nitrogen fertilizers. This has been encouraged by the demonstration of a marked nitrogen × phosphorus response for many species and by the significant increase in yield obtained with the use of nitrogen only

TABLE 4-13
The effect of time after fertilization on the sulphur content of *P. radiata* foliage treated with various fertilizers

Location	Fertilizer treatment	S sorption maximum of soil (μg/g)	Mean S (ppm) (as sulphate-S)										
			Years after planting										
			0.5	1	2	3	4	5	6	7	8	9	10
Buccleuch	N-P-K (25 g S/tree)	0.21	4	57	6								
	Elemental S (25 g S/tree)		1	76	18								
Mullions Range Cpt. 40	Super (500 kg/ha)	0.38			298	165	54	40	17	13	14		
	Super (500 kg/ha) Gypsum (1750 kg/ha)				371	384	186	174	295*	144	180		
Mullions Range Cpt. 30	Super (500 kg/ha)	0.38		668	226	173	65						
	Super (500 kg/ha) Ammonium sulphate (140 kg/ha)			298	188	161	71						
Mellong Range	Super (230 g/tree)	2.13				80		46	443**		167		
	Super lime (345 g/tree)					112		9	724**		208		
Belanglo	Super (500 kg/ha)†	4.44							315	174	192	152	60
	Ammonium phosphate (300 kg/ha)†								331	203	200	206	134
Penrose (second rotation)	Super (1000 kg/ha)††	4.42			254		235						
	Rock phosphate (500 kg/ha)††				225		268						
Wingello	Rock phosphate + S (280 g/tree)	57.06				259				264			
	Superphosphate (450 g/tree)					239				216			

* Super (750 kg/ha) applied five years after planting.
** Super (230 g/tree) applied five years after planting.
† Applied five years after planting. Lime super (300 g/tree) applied at planting.
†† Applied fifteen years prior to planting.

TABLE 4-14

Results for sulphate-S, and total-N in the foliage of *P. radiata* at Mullions Range S.F. for various fertilizer treatments and the related mensuration data

	Years after planting and fertilization																			
	2	3	4	5	6	7	8	2	3	4	5	6	7	8	2	3	4	5	6	7
Fertilizer Treatment	Mean S (ppm) (as sulphate-S)							Mean total N (%)							Mean plot height (m)					
Super (500 kg/ha) applied per tree	298	165	54	40	17	13	14	1.28	1.35	1.28	1.25	1.12	1.15	1.16	0.76	1.34	2.32	3.11	3.87	4.45
Super (500 kg/ha) applied per tree Lime (1,000 kg/ha) broadcast	167	143	41	67	19	11	15	1.28	1.26	1.33	1.26	1.08	1.10	1.21	0.85	1.53	2.71	3.66	4.58	5.28
Super (500 kg/ha) applied per tree Lime (1,000 kg/ha) broadcast Urea (56 g/tree)	51	163	22	77	13	4	12	1.26	1.38	1.48	1.38	1.13	1.11	1.18	0.92	1.53	2.78	3.60	4.45	5.06
Super (500 kg/ha) applied per tree Gypsum (1,750 kg/ha) broadcast	371	384	186	174	295*	144	180	1.33	1.33	1.35	1.25	1.21*	1.13	1.15	0.82	1.49	2.71	3.60	4.61*	5.64
LSD 0.05	160	120	67	34	120	72	122													

* Super (750 kg/ha) broadcast five years after planting.

TABLE 4-15
Sulphate-S and nitrogen foliar levels in *P. radiata* from various fertilizer treatments

Fertilizer	Mean foliar nutrient levels three years after planting and fertilization	
	Sulphate-S (ppm)	Total N (%)
Nil	59	1.32
Superphosphate (450 g/tree)	239	1.13
Rock phosphate (250 g/tree)	106	1.23
Rock phosphate + S (280 g/tree)	259	1.20
LSD (0.05)	120	

fertilizers on mature stands in Sweden and northwestern U.S.A. The northern hemisphere has generally much higher sulphur dioxide levels in the atmosphere than the southern and hence it would be expected that advantage could be taken of this assured supply of sulphur by adding nitrogen alone to the soil. But nevertheless the foliage from Douglas fir trees from two Washington forest areas yielded the following results (Table 4-16).

Similarly the sulphur status of foliage has been affected in an experiment in which nitrogen and phosphorus fertilizers have been used at planting with *P. radiata* (Table 4-17). Here the nitrogen only treatments gave nil response possibly due to the addition of nitrogen aggravating a situation where sulphur was already limiting. It is evident from these results that a serious reduction in sulphur reserves can be induced by using nitrogen sources which contain

TABLE 4-16
Effect of a nitrogen fertilizer on Douglas fir foliage

Fertilizer treatment	Foliar nutrient levels	
	% Nitrogen	Sulphate-S ppm S
Nil	1.34	760
	1.01	560
Nitrogen 450 kg/ha	2.50	0
	1.53	0

TABLE 4-17
Comparison of nitrogen and sulphur fertilizers on *P. radiata* and *P. elliottii*

Location	Fertilizer treatment	Ht (m) at age 5 yrs	Mean foliar nutrient levels	
			%N	Sulphate-S ppm S
Mullions Range (*P. radiata*)	Nil	0.49	1.99	28
	Urea (28 g/tree)	0.49	2.48	17
	Lime super (1,500 kg/ha)	2.71	1.28	167
Conjola (*P. elliottii*)	Nil	0.91	1.08	217
	$CaNH_4NO_3$	0.73	1.21	32
	Lime super (1,000 kg/ha)	4.51	0.98	110

no sulphur and that under circumstances similar to those which have been described, it can induce serious sulphur deficiency problems.

REFERENCES

(1) Afzal, M. (1972). BSc (hons) thesis, Australian National University.
(2) Beaton, J. D. (1966). *Soil Sci.*, **101**: 267.
(3) Hutton, J. T. (1958). *Arid Zone Research*, **11**: 285.
(4) Ingestad, T. (1963). *Medd. Skogsforskn. Inst. Stockholm*, **51**(7).
(5) Kelly, J. and Lambert, Marcia J. (1972). *Pl. Soil*, **37**: 395.
(6) Lambert, Marcia J. and Kelly, J. (1973). (unpublished data.)
(7) Lyle, E. S. Jr (1969). *Agron. J.*, **61**: 395.
(8) Truman, R. (1972). Forestry Commission of N.S.W., Tech. paper no. 19.
(9) Truman, R. and Turner, J. (1972). Forestry Commission of N.S.W., Tech. paper no. 18.

The Use of Sulphur and Its Compounds in Soil Amendment

J. LOVEDAY

Besides the correction of sulphur deficiency, there are two other purposes for adding sulphur, or its compounds, to soil. These are, firstly, to modify or maintain physical properties during reclamation and, secondly, to reduce pH and thereby to increase the availability of certain nutrients. This article will deal mainly with the former, but the latter use deserves brief mention. Indeed when reclaiming many saline soils both processes are involved concomitantly.

Soils may need reclaiming for a variety of reasons, but those which concern us here are salinity and high pH and undesirable physical characteristics which impede infiltration, root development and seedling emergence, for example.

Salinity, usually due to soluble sodium salts, is reduced by leaching, and if the soil contains sufficient soluble calcium to replace sodium on the exchange complex, then leaching in association with appropriate drainage may suffice for reclamation (22). In the absence of a source of calcium, leaching of soluble sodium salts leaves a legacy of exchangeable sodium on the exchange complex, i.e., a condition of sodicity. If sufficient swelling clay is present, once the electrolyte level in the soil solution is reduced by leaching, interlayer swelling increases, and dispersion and particle movement may be initiated (30,42). Both these processes contribute to the closing of conducting pores and hence to a reduction in hydraulic conductivity, the so-called 'freezing up' of a soil.

Loss of permeability of surface horizons by these processes may effectively prevent further leaching of soluble salts from the soil profile. However, for calcareous soils, though the permeability may be severely reduced initially, continued slow leaching over long periods allows solution of calcium carbonate which eventually achieves reclamation (6,22). Any added amendment which increases the solubility of calcium carbonate, e.g., acids or acid formers such as sulphur, can considerably hasten reclamation (23).

Many soils, e.g., in the Riverina of N.S.W., are naturally in a state of sodicity, without being excessively saline, at least not in the surface horizons. Presumably they have passed through the initial stages of leaching. Their permeability is intrinsically low, so that it is not a question of maintenance of permeability during leaching reclamation as it is with saline soils, but one of ameliorating an existing undesirable physical condition.

In either the saline or the sodic case the desired end product of complete reclamation is a calcium-dominated exchange complex, with sodium present in minor amounts. The level of salinity or sodicity which can be tolerated will vary with the plant species (40), and with other associated soil properties, e.g., clay content, clay type and degree of aggregate stability.

As indicated above the source of calcium may be the soil itself, but if in the form of calcium carbonate, the need for an acid or acid-forming amendment is likely. Alternatively, a source of calcium must be supplied, and of these available gypsum is the most widely used throughout the world (4,54). It is used almost exclusively in Australia.

Gypsum, or other compounds, may be applied directly to the soil, or dissolved in irrigation water. Sometimes in the latter case gypsum is dissolved to correct an unfavourable ionic balance in the water, which otherwise would lead to development of sodic soil conditions. The high-salt water method for reclaiming sodic soils (47) may involve the use of sea-water saturated with gypsum.

USE OF GYPSUM IN RECLAMATION IN AUSTRALIA

The value of gypsum has been known to farmers, e.g., in Hungary, for several centuries, but its use in Australia may only date from the early 1920s. At that time it was being tried on wheat-growing soils in Victoria (50) and on orchard soils in New South Wales (49) and South Australia. Since then, gypsum has been applied to a range of soils under varying circumstances and with varying success.

In dry-land farming the greatest use has probably been on wheat-growing soils of the Wimmera district, Victoria, where proximity to natural lake-bed deposits makes gypsum application economical (50). The soils to which application has been made include red-brown earths with shallow loam or sandy loam surface horizons over clay subsoils, grey clays which crack and set hard and which may or may not show some gilgai microrelief, and soloths with fine sandy loam surface horizons, including a bleached A_2, over heavy clay subsoils. Improved wheat yields are considered to be due mainly to prevention of crusting giving better emergence, and increased water entry into the soil during both the fallow and cropping phase (D. R. Rooney, priv. comm.). Residual effects from dressings of 3.75 t/ha probably last five to ten years. Again for wheat growing in the Dalby district, Queensland, trials on a grey clay have given yield responses to gypsum, increased emergence and improved physical condition of the soil (W. Bott, priv. comm.).

Favourable, but uneconomic, emergence and infiltration responses to gypsum have been recorded in reclaiming 'scalds', areas of erosion-exposed B horizons of red-brown earths, in both Victoria and New South Wales (20,21). The effect is especially apparent when light rains, sufficient to cause crusting in the absence of gypsum, are followed by dry conditions.

Under irrigation in the Riverina of New South Wales and Victoria, the most responsive soils have proved to be the hard-setting grey and brown clays, and shallow or transitional red-brown earths in which normal cultivation procedures enrich the surface soil with subsoil clay (28,33). Improved pasture emergence (10), rice seedling establishment (5), pasture and wool production (32), production of wheat (J. C. Noble and L. F. Myers, priv. comm.), of grain sorghum (34), and of cotton (31) have all been recorded on these soils following gypsum application. Of course, whether the result is attractive economically depends on the currently existing cost structure (44).

In a red-brown earth with about 18 cm of sandy loam surface, increased subsoil moisture was measured following surface application of gypsum (9). Pasture growth was significantly improved in this instance, but from other unpublished experiments it seems that significant responses have not usually been obtained on red-brown earths with deeper loamy surface horizons.

In the cases quoted above soils of intrinsically low permeability have responded to gypsum. But in Australia gypsum has not found such widespread use for more permeable saline soils, for which it might be expected to maintain permeability during leaching. The desalting of a saline Kerang clay (14) was successfully achieved by ponding and pumping from an underlying aquifer, and gypsum application did not increase the efficiency of the process. However, crusting and emergence difficulties, which developed after desalination, are responsive to gypsum application (F. Kaddous, priv. comm.). With respect to the timing of the gypsum application most experience is that it is more effective if added before cultivation and draining, rather than after desalting has been attempted (3,55,56).

Several leaching experiments on irrigated calcareous solonized brown soils (mallisols) in South Australia (W. B. Harris, priv. comm.; J. B. Robinson, priv. comm.) have failed to show any additional response to gypsum over and above that due merely to drainage and leaching. These results confirm earlier work of Thomas (52) on a Victorian mallisol and, presumably, may be explained by the presence of a sufficiently soluble source of calcium. A similar lack of gypsum response on a saline-alkali calcareous soil in the upper Colorado River Basin has been attributed to the presence of gypsum in the profile below 20 cm (2).

In many parts of the world gypsum and acidulents find widespread use, in association with drainage, for reclaiming saline soils containing sodium carbonate (1,19,22). In such soils very high levels of ESP* develop because the equilibrium reaction

$$Na_2\,CO_3 + Ca^{++}\,\text{clay} \rightleftharpoons Na_2^{+}\,\text{clay} + CaCO_3$$

moves continually to the right as calcium carbonate is precipitated (22).

MODIFICATION OF SOIL PROPERTIES BY GYPSUM

Addition of gypsum, which dissolves slowly, increases the electrolyte content of the soil solution for a time and replaces sodium on the exchange complex by calcium. Because there is reduced dispersion and interlayer swelling of the clay particles the structural pore space, once formed, is maintained (28,46) and hydraulic conductivity increased markedly (45). These are the basic effects underlying the modifications to soil tilth, crust strength, water infiltration and storage, drainage and leaching, all of which have been recorded on gypsum-responsive soils.

---

* Exchangeable sodium percentage.

Soil tilth and crust strength

The surface condition of Australian soils of ESP > 5 or 6 (sodic soils (35)) frequently tends to be massive or coarse blocky, breaking with difficulty to aggregates fine enough for a seed bed. Wetting by rain or low electrolyte irrigation water causes dispersion and, as the soil dries, a hard surface crust forms. The range of water contents over which cultivation can be carried out is narrow.

There are frequent records of gypsum-induced improvements in the ease and effectiveness of tillage for producing a seed-bed and for weed control, the range of water content for tillage and its timeliness (e.g. (28,49), Rooney, priv. comm.). But no attempts appear to have been made to assess the value of reduction in tillage and land grading costs. Perhaps no direct economic value can be placed on effects such as greater flexibility in the date of sowing and harvesting and the more even irrigation resulting from better grading, yet these can be of considerable practical advantage.

The strength of a surface crust is very largely dependent upon its moisture content and, because gypsum treatment influences the rate of surface drying (28), the rate of development of strength, as well as the final strength reached, may be affected. These effects have usually been recorded in terms of marked improvements in seedling emergence and establishment (5,10,28), but they have also been measured as reductions in the modulus of rupture (15) and in resistance to a penetrometer. The force required to push a penetrometer point (3 mm diameter, 60° cone) 1 cm into soil which has been allowed to air-dry following a single flooding of a cultivated seed-bed was measured in three experiments (Loveday, unpublished data). The differences between the experiments (Table 4-18) have been mainly attributed to the degree of wetting and the subsequent rates of drying. Even though 15 of the 53 soils (representative of Riverina and Namoi irrigation areas, N.S.W.) did not respond to gypsum, the overall means still show very substantial effects.

Water infiltration and storage

Water entry and storage benefits due to gypsum have been described for a variety of soils throughout the world, e.g., for a grey cracking clay in the Sudan Gezira (16), for a fine sandy loam in California (39), for a sandy loam over clay subsoil in New South Wales (9) and for grey and brown sodic clays, also in New South Wales (8,27).

The improvement in water storage and the depth of soil affected are dependent upon the nature of the soil and pre-existing moisture conditions. For

TABLE 4-18
Effect of gypsum on penetrometer resistance (kg), (the values are means for 53 soils, 30 readings per soil)

Treatment	Experiment No. 1	2	3
Nil gypsum	3.6	11.6	10.1
Gypsum (7.5 t/ha)	1.7	2.8	2.0

a brown sodic clay (Loveday, unpublished data) total storage was increased two and a half times by applied gypsum. However, for the sandy loam over clay mentioned above (9) only a small non-significant increase occurred in the subsoil 20-50 cm, though both pre- and post-irrigation water contents were higher due to gypsum.

Effects on water storage obviously result from greater infiltration and profile permeability. Practical difficulties have frequently discouraged attempts at field measurement of infiltration rate on cracking clay soils. For a fine sandy loam (39) the rate was approximately doubled by gypsum. Laboratory determined hydraulic conductivity data for sodic soils usually demonstrate substantial effects of gypsum (8,9,27).

Drainage and leaching

Reference has already been made to the widespread use of gypsum, in association with drainage, to reclaim saline soils dominated by sodium carbonate. Resulting profile changes in soluble constituents, exchangeable sodium and pH have been documented for various soils (e.g., (19,23)).

Though the prime object of applying gypsum to many sodic clays is to ameliorate their undesirable physical properties, a valuable side effect which may result is extra drainage and attendant leaching of salts to layers beneath the root zone of crops. Three seasons of irrigation following gypsum application to a brown sodic clay, reduced chloride content in the profile to 125 cm depth and increased it between 125 and 275 cm. Comparable figures in the absence of gypsum were 85 and 200 cm (26).

PROFITABILITY OF GYPSUM APPLICATION—RESIDUAL EFFECTS

Where gypsum is used in association with artificial drainage to permanently reclaim an unproductive saline soil, its residual effects on future crop production clearly can be very considerable.

However, in Australia, gypsum is often applied to sodic or saline-sodic soils in situations where artificial drainage is not possible. Moreover, the amount applied may be sufficient to ameliorate only the upper 20-30 cm of soil. In these circumstances the question of profitability arises and it is usually put in the form of 'how large is the increase in production and for how long does it persist?' Whether production is increased or not may depend largely on the interaction with seasonal conditions, an interaction which can be highly significant when properties concerned with soil water status are involved. For dry-land wheat cropping in a season of inadequate rainfall in the latter part, greater plant density resulting from gypsum could perhaps lead to premature 'haying-off' and even yield reductions. On the other hand, for summer crops under irrigated conditions, the greatest yield increases due to gypsum may occur in drier seasons when increased infiltration and water storage become effective. It seems likely that seasonal interactions have been responsible for variable results frequently obtained in field experiments designed to measure gypsum response in terms of plant production. Those gypsum-induced modifications to soil properties which directly affect costs, e.g., reduced tillage requirement for land grading and seed-bed preparation, should be taken into account when determining profitability.

Despite the problems with seasonal variation, the economics of gypsum application have been claimed to be satisfactory under a variety of conditions (e.g., (32,50)). Residual effects, important in this regard, have been noted for eight to ten years under dry-land farming (50) and at least three to four years under irrigation (32,48).

GYPSUM REQUIREMENT AND RECOGNITION OF GYPSUM-RESPONSIVE SOILS

The established methods for determining a soil's 'gypsum requirement' (40) are firstly, to calculate the gypsum required to replace exchangeable sodium and secondly, to determine the amount of calcium lost by a saturated gypsum solution when shaken with the soil sample. A similar method (43) uses the measurement of 'calcium deficit', defined as the amount of calcium needed to saturate the soil with divalent cations.

Recent computer-based methods (11, 51) call for more detailed soil and water analyses to take account of the multiplicity of chemical reactions occurring in natural soils and to deal with stratified soils. Comparison of predicted and field results have been encouraging.

These computer methods clearly make a considerable advance in allowing the fusing together of detailed knowledge of soil processes in reclamation. However, saline and sodic soils are noted for their spatial variability and this taken together with the amount of analytical detail required may limit their general and widespread use.

The foregoing methods involve the exchange reaction of calcium for sodium and aim for either complete replacement of sodium or replacement to some arbitrary level over an arbitrarily specified depth. They ignore the possibility of worthwhile responses from the electrolyte effect (38). The dissolved gypsum treatment advocated for pasture establishment on Australian sodic clays (10) is based on this electrolyte effect; the more or less complete replacement of exchangeable sodium by calcium is not an objective of the treatment. Likewise much of the initial response when solid gypsum is broadcast can be attributed to the electrolyte effect while gypsum is still dissolving (17).

Thus a reliable method for simply recognizing gypsum-responsive soils, without attempting to nominate a 'gypsum requirement' to cause permanent changes in ESP, could be of value. Work done with this end in view (Loveday, unpublished data), shows that two easily determined properties—dispersion index (25) and pH—can be used to recognize gypsum-responsive soils with reasonable confidence.

USE OF SULPHUR, SULPHURIC ACID AND OTHER ACIDULENTS —COMPARATIVE TRIALS WITH GYPSUM

As indicated previously, acid or acid forming amendments are of value for physical amelioration for those soils which contain calcium carbonate able to be dissolved to release calcium. During the process pH is lowered with the well-known effects on the availability of zinc, manganese and iron. Higher uptake of phosphorus by rye grass has also been reported when a dressing of sulphur lowered pH from 8.8 to 6.1 (18). The possibility exists then, for the use of acidulents simply to increase nutrient availability. In this context there appears to be a need for a clearer definition of the soil conditions worth treating and more consideration of such aspects as placement in relation to root development.

Acidulents have not been widely used in Australia. On calcareous solonized brown soils at Mildura, Victoria, a residue from gas works (containing about 30 per cent sulphur) was commonly applied to chlorotic vines (J. G. Baldwin, priv. comm.) but the practice has apparently been discontinued. An instance of enhanced manganese availability from a sulphur dressing has been reported for a sandy loam soil (pH 6.7) on limestone from Flinders Island (D. Wright, priv. comm.).

Reports from elsewhere, comparing amendments, suggest that their relative effectiveness varies from place to place. An alkali Fresno soil, when treated with equivalent amounts of gypsum, sulphuric acid and sulphur, responded best, in terms of pasture yield, to sulphuric acid; gypsum gave some initial response, whereas sulphur treated plots were no better than the control. Gypsum and sulphuric acid produced similar marked reductions in soluble and exchangeable sodium levels (36). In contrast, sulphur was more effective than gypsum in treating 'slick spots' (13), by producing better aggregation, higher water intake and deeper root penetration of alfalfa.

Again in Azerbaidzhan, saline-sodic soils have responded better to 'gaga', a natural gypsum deposit, than to sulphuric acid or burnt pyrite (1), whereas in Armenia on similar soils, sulphuric acid is said to produce better reclamation than any of a great range of amendments including gypsum and sulphur (37). In this case the acid is added to water in leaching ponds, the soil of which has previously been tilled to 40-80 cm.

Calcium and ammonium polysulphides, which oxidize readily in the soil, have been used to some extent in the U.S.A. (29). When applied in amounts equivalent to the 'gypsum requirement' they increase permeability, but at the lower rates (e.g., 2×10^3 1/ha) usually recommended commercially they may be ineffective (29). In a comparison between calcium and ammonium polysulphides and sulphuric acid in the Imperial Valley, California (41), only ammonium polysulphide increased water intake rates. Again, for three San Joaquin Valley soils both gypsum and sulphur dioxide improved permeability to low electrolyte waters, but calcium polysulphide was ineffective (12). The value of the polysulphides thus seems to be equivocal.

In some localities cheap acidic industrial wastes may be available, e.g., acid resins from the oil industry, and these may be applied with or without ground limestone depending on the calcium status of the soil (7).

The more widespread testing and use of some of these products may require development of special handling techniques. Waste sulphuric acid and sulphur dioxide, for instance, cannot be added to water which is to pass through metal or concrete pipes. Commercial use of sulphuric acid nevertheless has been reported from Colorado where it was chiselled into the soil in bands 45 cm apart (54). The dust problem, which has proved a serious disadvantage to the use of elemental sulphur, seems likely to be overcome by using conventional fluid fertilizer equipment to apply water suspensions containing 55 to 60 per cent sulphur (53). Wider use of dissolved instead of broadcast gypsum has been made possible by the development of a special machine for its dissolution (24).

REFERENCES

(1) Agaev, B. M. (1965). *Agrokem. Talajt.*, **14** (suppl.): 189.
(2) Amemiya, M., Robinson, C. W. and Cowley, E. W. (1956). *Proc. Soil Sci. Soc. Am.*, **20**: 423.
(3) Anderson, G. D. (1966). *E. Afr. Agric. For. J.*, **31**: 424.
(4) Artjushin, A. M., Derzhavin, L. M., Krepak, K. A., Pak, K. P. and Yakovleva, M. E. (1969). *Agrokem. Talajt.*, **18** (suppl.): 307.
(5) Barley, K. P. and Hutton, J. T. (1956). *Aust. J. Agric. Res.*, **7**: 110.
(6) Bower, C. A. and Goertzen, J. O. (1958). *Proc. Soil Sci. Soc. Am.*, **22**: 33.
(7) Bocksai, J. (1969). *Agrokem. Talajt.*, **18** (suppl.): 336.
(8) Bridge, B. J. (1968). *Soil Physical Properties as Influenced by Gypsum.* MSc (Ag.) thesis. University of Sydney.
(9) Bridge, B. J. and Kleinig, C. R. (1968). *Trans. IX Int. Congr. Soil Sci.*, Adelaide, **1**: 313.
(10) Davidson, J. L. and Quirk, J. P. (1961). *Aust. J. Agric. Res.*, **12**: 100.
(11) Dutt, G. R., Terkeltoub, R. W. and Rauschkolb, R. S. (1972). *Soil Sci.*, **114**: 93.
(12) El Tayib, Y. M. (1972). *Effect of Various Sulphur Compounds on Infiltration Rates of Selected Soils.* PhD thesis, University of California.
(13) Fitts, J. W., Lyons, E. S. and Rhoades, H. F. (1944). *Proc. Soil Sci. Soc. Am.*, **8**: 432.
(14) Garland, K. R. and Jones, G. O. (1963). *Aust. J. Exp. Agric. Anim. Husb.*, **3**: 177.
(15) Graveland, D. N. and Toogood, J. A. (1963). *Can. J. Soil Sci.*, **43**: 1.
(16) Greene, H. (1928). *J. Agric. Sci. Camb.*, **18**: 531.
(17) Greene, H. and Snow, O. W. (1939). *J. Agric. Sci. Camb.*, **29**: 1.
(18) Gupta, M. B. S. and Cornfield, A. H. (1964). *Pl. Soil*, **21**: 388.
(19) Herke, S. and Harmati, I. (1965). *Agrokem. Talajt.*, **14** (suppl.): 313.
(20) Jones, R. M. (1962). *J. Soil Conserv. Serv. N.S.W.*, **18**: 78.
(21) Jones, R. M. (1969). *J. Soil Conserv. Serv. N.S.W.*, **25**: 104.
(22) Kelley, W. P. (1951). *Alkali Soils, their Formation, Properties and Reclamation.* Reinhold, New York.
(23) Kelley, W. P. and Arany, A. (1928). *Hilgardia*, **3**: 393.
(24) Kelsall, D. F., Crockford, R. H. and Davey, E. T. (1960). *J. Agric. Engng Res.*, **5**: 355.
(25) Loveday, J. and Pyle, J. (1973). Tech. Pap. Div. Soils CSIRO Aust. No. 15.
(26) Loveday, J., Muirhead, W. A. and Gunn, A. H. (1974). 'The effect of gypsum and tillage on leaching of an irrigated saline-sodic clay', *Trans. X Int. Congr. Soil Sci.*, Moscow, in press.
(27) Loveday, J., Saunt, J. E., Fleming, P. M. and Muirhead, W. A. (1970). *Aust. J. Exp. Agric. Anim. Husb.*, **10**: 36.
(28) Loveday, J. and Scotter, D. R. (1966). *Aust. J. Soil Res.*, **4**: 55.
(29) McGeorge, W. T., Breazeale, E. L. and Abbott, J. L. (1956). Ariz. Agric. Exp. Sta. Tech. Bull. 131.
(30) McNeal, B. L., Norvell, W. A. and Coleman, N. T. (1966). *Proc. Soil Sci. Soc. Am.*, **30**: 313.
(31) Muirhead, W. A., Loveday, J. and Saunt, J. E. (1970). *Aust. J. Exp. Agric. Anim. Husb.*, **10**: 325.
(32) Myers, L. F. and Squires, V. R. (1970). *Aust. J. Exp. Agric. Anim. Husb.*, **10**: 36.
(33) Noble, J. C. (1963). N.S.W. Dep. Agric. Div. Pl. Ind., 9pp.
(34) Noble, J. C. and Kleinig, C. R. (1971). *Aust. J. Exp. Agric. Anim. Husb.*, **11**: 53.
(35) Northcote, K. H. and Skene, J. K. M. (1972). Soil Publ. CSIRO Aust., No. 27.
(36) Overstreet, R., Martin, J. C. and King, H. M. (1951). *Hilgardia.*, **21**: 113.
(37) Petrosian, G. P. and Tchitchian, A. I. (1969). *Agrokem. Talajt.*, **18** (suppl.): 121.
(38) Quirk, J. P. and Schofield, R. K. (1955). *J. Soil Sci.*, **6**: 164.
(39) Reitemeier, R. F., Christiansen, J. E., Moore, R. E. and Aldrich, W. A. (1948). USDA Tech. Bull. No. 937.

(40) Richards, L. A. (ed.) (1954). USDA Handbk No. 60.
(41) Robinson, F. E., Cudney, D. W. and Jones, J. P. (1968). *Calif. Agric.*, **22**: 10.
(42) Rowell, D. L., Payne, D. and Ahmad, N. (1969). *J. Soil Sci.*, **20**: 176.
(43) Schofield, R. K. and Taylor, A. W. (1961). *J. Soil Sci.*, **12**: 269.
(44) Scott, B. J., Evans, L. H. and Noble, J. C. (1968). *Agric. Gaz. N.S.W.*, **79**: 493.
(45) Scotter, D. R. and Loveday, J. (1966). *Aust. J. Soil Res.*, **4**: 69.
(46) Sedgley, R. H. (1962). *Soil Sci.*, **94**: 357.
(47) Shah Muhammed, McNeal, B. L., Bower, C. A. and Pratt, P. F. (1969). *Soil Sci.*, **108**: 249.
(48) Sharma, M. L. (1971). *Aust. J. Soil Res.*, **9**: 73.
(49) Shepherd, A. N. (1925). *Agric. Gaz. N.S.W.*, **36**: 261.
(50) Sims, H. J. and Rooney, D. R. (1965). *J. Dep. Agric. Vict.*, **63**: 401.
(51) Tanji, K. K., Doneen, L. D., Ferry, G. V. and Ayers, R. S. (1972). *Proc. Soil Sci. Soc. Am.*, **36**: 127.
(52) Thomas, J. E. (1939). Bull. CSIR Aust. No. 128.
(53) Thorup, J. T. (1972). *Sulphur Inst. J.*, **8**: 16.
(54) Tisdale, S. L. (1970). *Sulphur Inst. J.*, **6**: 2.
(55) Van Beekom, C. W. C., Van Den Berg, C., De Boer, Th.A., Van Der Molen, W. H., Verhoeven, B., Westerhof, J. J. and Zuur, A. J. (1953). *Neth. J. Agric. Sci.*, **1**: 153.
(56) Verhoeven, B. (1965). *Agrokem. Talajt.*, **14** (suppl.): 263.

5 Diagnosis and Prediction of Sulphur Deficiency

Plant Assays for the Recognition of Sulphur Deficiency

M. B. JONES

The ultimate in assaying the sulphur status of a crop might be a small hand-held instrument that can test the plants for S in the field. Although we cannot now gain information this fast, we can within a few days of sampling determine total S, soluble or reducible S, N:S ratios, and certain biochemical compounds, or make a physiological assessment of a crop. These tests have all given satisfactory indices of sulphur status under field conditions, although the delay between sampling and the analysis report, often may limit its usefulness to recommendations for the following year.

Even with the best test information readily available, knowledge of the crop environment is essential for proper evaluation of the test data. For example, the following factors may influence test results: the plant itself (species, age and part analyzed), and environmental conditions (deficiencies or toxicities of other mineral elements, drought, high or low temperature, disease, plant defoliation and type of crop culture, i.e., field *v.* potted soil *v.* nutrient solution). In fact, any environmental factor that limits plant growth more than sulphur deficiency will probably increase the concentration of sulphur compounds in the plant. Usually lower critical concentrations have been obtained in pot and nutrient solution experiments than in the field, the critical concentration being defined as that concentration of sulphur or sulphur compound required to give a yield of 90 to 95 per cent of maximum.

Procedures for taking and handling the samples can also influence results. The sample should therefore be representative of the area to be tested and handled so as to avoid spoilage or contamination.

Visual symptoms can be helpful when deficiencies are severe, but the same factors listed above make interpretation difficult. Also, sulphur deficiencies may reduce yields without any symptoms showing on the plant. Where pasture plants such as subclover (*Trifolium subterraneum* L.), soft chess (*Bromus mollis*) and ripgut (*B. rigidus*) are growing in competition with one another, a reduction in clover percentage can indicate low S (7). When Hardinggrass (*Phalaris tuberosa* var. stenoptera) occurs in such a field, more vigorous clover growing in clumps of Hardinggrass is indicative of sulphur deficiency since Hardinggrass accumulates sulphur. Any visual symptoms in the field must be confirmed by chemical tests or fertilizer trial.

Total sulphur

The total sulphur concentration in plant tissue has been widely used as a

TABLE 5-1

Effect of S fertilization on concentration of S and reducible S in pot-grown annual species

Days after planting	S applied	Subclover	Soft chess	Ripgut	Medusahead	Filaree	Total yield
	kg/ha			% S			g/pot
92	0	.105a*	.157a-c	.126ab	.150a-c	.114a	6.9a
	6	.182b-e	.176a-d	.223d-g	.176b-d	.191c-f	9.1b
	11	.251f-h	.278g-i	.232d-g	.192c-f	.243e-h	9.8b
	22	.341ij	.396j	.235d-g	.259gh	.253f-h	8.3b
	34	.396j	.390j	.252f-h	.250f-h	.301hi	9.2b
155	0	.051ab	.065b-d	.047ab	.062b-d	.038a	22.0c
	6	.051ab	.083e-g	.054a-c	.075d-g	.050ab	35.4d
	11	.067b-d	.106hi	.072c-f	.076d-g	.063b-d	40.7e
	22	.133jk	.136k	.089f-h	.113ij	.082d-g	44.4f
	34	.161l	.142kl	.095g-i	.109i	.083e-g	45.0f
				ppm reducible S			
92	0	100a	100a	70a	100a	90a	
	6	820d	570c	420bc	280b	670cd	
	11	1320f	1320f	870de	610d	1120ef	
	22	2690h	1970g	1250ef	1020ef	1390fg	
	34	2760h	2140g	1280ef	1030ef	1520fg	
155	0	40i	110a-j	40i	80a	51i	
	6	50i	240b	80a	140j	80a	
	11	120j	410c	270b	200b	220b	
	22	490c	700d	360b	390b	290b	
	34	820d	800d	490c	380b	280b	

* Values followed by the same letter are not significantly different at the 5 per cent level.

measure of sulphur status and now is favoured by many as the most suitable index because it is relatively easy to determine. The concentration required must be determined for individual species, plant parts, and growing conditions. For example, Eaton (4) listed sulphur values for twenty-six plant species, with lows ranging from 0.05 to 0.35 per cent S.

Differences in sulphur concentrations were not great among annual pasture plants competing for sulphur in the same pots and fertilized with 25 ppm N, 44 ppm P, and 55 ppm K (Table 5-1) (6). Some differences were significant, however, and depended on rate of S applied and date of sampling. On day 92 after planting, sulphur concentrations did not differ significantly among the five species growing in the check treatment. Where S at 34 kg/ha was applied, subclover was highest in S and Medusahead (*Elymus caput-medusa* L.) was lowest. On day 155, filaree (*Erodium botrys*) was lowest in S both where no S was applied and where 34 kg S/ha was applied. When sulphur percentages in each species were plotted against total yield the subclover appeared to be the best plant for assay purposes. Only in the check treatment was subclover S deficient on day 92, but by day 155 the three lowest S treatments were deficient. Thus, tissue S values evidenced S status strictly for the time of sampling, without indicating the future status.

A field study (Table 5-2) revealed fewer consistent differences among species, except that Hardinggrass accumulated more S than other species when S was applied.

The plant part sampled can make an important difference. Jones *et al.* (9) reported the critical S value to be 0.23 per cent S in flowering bur clover leaves but 0.08 per cent in stems, indicating that plant parts should not be mixed for analysis.

Age of the plant or the plant part often markedly affects the level of sulphur found. Eaton (4) concluded that no general statement could be made since sulphur levels increased with age in some plants but decreased in others. The change with time also depended on the nutritional status of the plant. In any case the behaviour must be determined for each species.

The availability of other nutrients can have an important effect on sulphur concentrations in plant tissue. Table 5-3 illustrates how increased phosphorus in a pot study reduced the percentage of sulphur in the plant (8). From a field study, Bouma *et al.* (2) reported that adequate P status greatly improved the relationship of yield to S concentration in subclover over that where no P was applied.

Sulphur concentrations in annual clovers sampled by Martin and other California workers during years of spring drought were higher than expected (unpublished). In another unpublished study, subclover leaves of the same age had higher S concentrations if the plants had been defoliated several times.

TABLE 5-2

Effect of S fertilization on total S, reducible S and N:S ratios of field-grown annual grassland species

S applied	Subclover	Rose clover	Native clovers	Annual grasses	Harding-grass	Total yield kg/ha
kg/ha			% S			
0	.081ab*	.055a	.083ab	.081ab	.088ab	4020a
22	.096a-c	.085ab	.094a-c	.099bd	.166e	6730b
45	.137de	.092a-c	.154e	.146de	.247f	8100c
90	.157e	.149e	.130cd	.174e	.422g	8150c
			ppm reducible S			
0	70a-c	30a	80bc	130de	210e-g	
22	150de	70a-c	290fg	340g	870j	
45	280fg	150de	430h	550h	1570i	
90	490h	360g	510h	700ij	3160a	
			N:S			
0	28.2f	35.8g	26.9f	14.1cd	18.8de	
22	26.8f	25.2f	27.3f	14.0cd	9.0a-c	
45	18.9de	25.5f	18.9e	10.5bc	7.3ab	
90	18.6de	16.5de	22.6ef	10.2bc	4.0a	

* Means followed by the same letter are not significantly different at the 5 per cent level.

Reducible sulphur

Sulphate-sulphur concentrations in plant tissue have also been widely used as a measure of sulphur status. The term 'sulphate' has been used rather loosely. Spencer (pers. comm.) suggested using the term 'reducible sulphur' to refer to that fraction of plant S reducible by HI when introduced (as dry plant material) directly into the digestion flask. The term 'soluble S' was used to describe S from plant material extracted with 2 per cent acetic acid (5) and then reduced with HI; the latter method gave results similar to those with the direct reduction method.

There are theoretical reasons for using reducible or soluble sulphur in diagnosis. Plants absorb S in the form of sulphate, which is reduced to other forms in plant metabolism. When S is in short supply, most of the S is incorporated into protein and little sulphate can be detected in the plants; when S is present in excess, however, sulphate accumulates in the plant. The coefficient of variation is usually higher for reducible S values than for total S.

While reducible S is generally quite low in most plants, critical levels should be considered on the basis of individual plant species. Examples of reducible S in five species competing in pots (Table 5-1) indicates that species differ significantly in changes with age and level of applied S. Samples taken from the field (Table 5-2) also showed differences between species. The legumes gave a clearer break between deficient and adequate sulphur levels than did grasses when reducible sulphur was plotted against forage yields. In bur clover, differences in reducible S levels in plant parts were not nearly as great as differences in total S (9). For annual pasture legumes, whole plant

TABLE 5-3
Effect of P and S interactions on subclover yields, S concentrations, and N:S ratios

S applied	P applied mg/pot			
	0	50	100	150
mg/pot		Yield, g per pot		
0	2.2a*	2.9a	3.4a	2.2a
10	2.0a	10.7bc	12.4cde	11.6bcd
20	2.3a	11.8b-e	14.8fg	13.6ef
30	2.3a	10.5b	12.9de	15.9g
		% S		
0	.060c	.035a	.037ab	.036ab
10	.153fg	.053b	.052abc	.058bc
20	.168gh	.094de	.084d	.090de
30	.188h	.132f	.108e	.107de
		N:S		
0	27b	42a	37a	38a
10	13d	29b	29b	26bc
20	12d	18d	19cd	19cd
30	11d	12d	16d	15d

* Means followed by the same letter are not significantly different at the 5 per cent level.

tops are as good as any selected part. Until proved otherwise for each plant species, however, it should be assumed that plant part does make a difference.

The level of other nutrients affects the reducible S values in essentially the same way as total S. When legumes and grasses are growing in a mixed stand, reducible S in the legumes usually gives a better picture of S status than reducible S in the grasses. Grasses are usually nitrogen-deficient and therefore accumulate higher levels of reducible S. When the grasses had adequate N, reducible S in soft chess was equivalent to that in subclover (Table 5-4 (7)). Jones and Martin (7) also found that the level of reducible S increased when subclover wilted and began drying at the end of the growing season.

The reducible S level in subclover tops can give an indication of S status at the moment of sampling. If values are above the critical level when the plants are young, however, it is difficult if not impossible to tell whether the S will be sufficient for optimum growth through the entire life of the plant.

Nitrogen:sulphur ratios

Dijkshoorn and Van Wijk (3) proposed using the ratio of N to S in plant material (17 for legumes and 14 for grasses) as a guide to diagnosing S deficiency on the basis that any less S in the plant than that needed to maintain the ratio of N to S in protein would constitute an S deficiency suppressing growth. Pumphrey and Moore (12) established a critical N:S ratio of 11 in alfalfa tops and indicated that N:S ratios were fully as good as S concentration for diagnosing S deficiency in alfalfa in the early bloom stage of growth; the N:S ratio was affected less by stage of growth than were S percentages, but analytical costs were doubled.

McNaught and During (11) indicated that N:S ratios greater than 16 in ryegrass tops and 18 in white clover leaves indicated S deficiency. Data of McNaught and Chrisstoffels (10) indicated that N:S ratios were lower in petioles than in leaflets of white clover. Alfalfa also had a lower N:S ratio in stems than in leaves, whereas there was no difference in red clover, indicating that plant parts must be tested for each species.

When N:S ratios were computed for mixed annual grassland species (Table

TABLE 5-4

Effect of increasing rates of S applied in October on reducible S concentrations in subclover and soft chess on the following 1 May. All plots were fertilized with 160 pounds N per acre in October

S applied	Subclover	Soft chess
kg/ha	ppm SO_4-S	
0	70a*	80a
22	180b	200b
45	380c	430c
90	590d	570d

* Means followed by the same letter are not significantly different from each other at the 5 per cent level.

5-2 (7)), subclover and rose clover gave values about twice that of soft chess, indicating higher S and lower protein levels in the grass than in the clover. In Table 5-2 N:S ratios can also be compared with reducible S values as criteria to evaluate S status. The two methods do not always agree. For example in the 90 kg S/ha treatment the N:S ratio in subclover is 18, indicating a borderline S deficiency, but the reducible S value of 490 ppm indicated adequate S. The N:S ratio in annual grasses (mostly soft chess) was 10, apparently indicating adequate S but also indicating that the grass was deficient in N. The N:S ratio of 14 in annual grasses where no S was applied did not reveal the S deficiencies.

Biochemical tests

Since sulphur is an essential component of protein, the ratio of protein to nonprotein nitrogen can be used to detect sulphur deficiency in plants. Randall (13) removed young leaves from plants grown at a range of sulphur levels and incubated them in solutions with and without sulphate. Levels of protein nitrogen remained steady or increased in sulphur-deficient leaves placed in sulphate solutions, while protein N in leaves supplied with adequate S decreased even where S was in the solution. The test is applicable only to young leaves still undergoing expansion. This test has the advantage of being specific in measuring the adequacy of S in the plant at sampling; however, it does require fresh plant material, incubation and extraction, and two N determinations.

Physiological assessment

Bouma and Dowling (1) compared the increase in plant growth as measured by leaf area after test plants were transferred from the field to nutrient solutions with and without S. Plants with adequate sulphur continued to grow at the same rate in both solutions, showing no difference in leaf area over a seven-day period, whereas plants growing at suboptimal sulphur levels showed smaller increases in leaf area when transferred to solutions lacking sulphur. The response pattern indicated only the presence or absence of stress from low sulphur.

CONCLUSIONS

To be useful to agriculture over a wide range of conditions plant assays for the recognition of sulphur deficiency must be fast and inexpensive. Sampling procedures must be simple enough that farmers can take them, send them to distant laboratories, and get back meaningful results. Running reducible S on dried plant material has the advantage of not requiring the separation of plant parts (such as stems and leaves) that is required with total S; however, reducible S values usually have higher coefficients of variation than total S values. The procedure used might well depend upon how the laboratory is equipped. The N:S ratios can be used when both N and S determinations are made.

Biochemical tests are useful in some special instances where testing facilities are close to the source of fresh plant material.

REFERENCES

(1) Bouma, D. and Dowling, E. J. (1966). *Aust. J. Agr. Res.*, **17**: 633.
(2) Bouma, D., Spencer, K. and Dowling, E. J. (1969). *Aust. J. Exp. Agric. Anim. Husb.*, **9**: 329.
(3) Dijkshoorn, W. and Van Wijk, A. L. (1967). *Pl. Soil*, **26**: 129.
(4) Eaton, F. M. (1966). *Sulfur-diagnostic Criteria for Plants and Soils.* University of California Press.
(5) Johnson, C. M. and Ulrich, A. (1959). Calif. Agr. Expt. Sta. Bull., **766**: 25.
(6) Jones, M. B. (1963). *Agron. J.*, **55**: 251.
(7) Jones, M. B. and Martin, W. E. (1964). *Proc. Soil Sci. Soc. Am.*, **28**: 539.
(8) Jones, M. B., Oh, J. H. and Ruckman, J. E. (1970). *N.Z. Grassld Assoc. Proc.*, **32**: 69.
(9) Jones, M. B., Ruckman, J. E. and Lawler, P. W. (1972). *Agron. J.*, **64**: 55.
(10) McNaught, K. J. and Chrisstoffels, P. J. E. (1961). *N.Z. J. Agric. Res.*, **4**: 177.
(11) McNaught, K. J. and During, C. (1970). *N.Z. J. Agric. Res.*, **13**: 567.
(12) Pumphrey, F. V. and Moore, D. P. (1965). *Agron. J.*, **57**: 364.
(13) Randall, P. J. (1970). *Proc. XI Int. Grassld Congr.*, p. 338.

Soil Assays for the Recognition of Sulphur Deficiency

H. M. REISENAUER

Sulphur deficiencies in plants are occurring with increasing frequency in the agricultural areas of the world (6,11). This is a result of practices which on the one hand have increased crop yields and S requirements and on the other have reduced additions of the element as constituents of N-P-K fertilizers, as pesticides, and from the atmosphere. Crop responses to S fertilization are not easily predicted. They do not follow any particular patterns of soils, climate or native vegetation (28). Their prediction is further complicated by the multiplicity of sources of the element to plants and the varying contributions from each source. Significant quantities may be supplied from the soil, from the atmosphere either in the precipitation or through direct absorption of gaseous forms, from the irrigation waters, and from fertilizers and pesticides. Any one of these sources may supply the total requirement of a crop. Soil tests give an estimate of the amount of the element that will be supplied by the soil and thus provide a part of the information essential to sound S fertilization recommendations.

SOIL SAMPLING

Self testing involves sampling, analysis, correlation and interpretation. Sampling may well be the factor most limiting the success of soil testing. The common assumption is that the analysis of the sample accurately reflects the status of the nutrient in the area from which the sample was collected. This assumption holds only if the area sampled is uniform, and if the sample is properly taken, at a meaningful time, and is correctly prepared.

No single philosophy or procedure for soil sampling is universally accepted. Most instructions suggest selection of the sampling area based on visual or known differences in soil, topography or past management (30,32,33). The size of the sampling area is normally restricted to a few hectares but guidelines vary. Commonly 15 to 30 subsamples of the plough-layer for cultivated crops, and of the top 7 to 10 cm for sod crops are suggested. Either random or systematic patterns of collecting subsamples may be recommended. Subsoil samples may or may not be required.

Our present sampling procedures are inadequate in several respects and need to be re-examined. Sampling plans designed to both represent an area, and to evaluate and outline soil variation are needed. They should allow for repeat samplings to estimate seasonal variation and long-term trends in fertility levels. And finally, more extensive use of other supportive information—detailed soil maps, aerial photographs, crop characteristics including tissue analysis, etc.—should be encouraged.

Soil sampling for sulphate status presents special problems. The extractable sulphate level of a soil varies with season, with depth in the profile, and with treatment of the sample. Seasonal differences result from varied rates of crop removal, of losses by leaching, and of additions from soil organic matter decomposition, from the atmosphere, irrigation water, etc. The distribution of extractable sulphate with depth in the profile varies greatly between soils. Accumulations may be found in either surface or subsurface layers. Since all but the organically combined and precipitated (as $BaSO_4$ or in calcareous horizons) sulphate of soils have high plant availability coefficients (4,10,46), crop rooting $\times$ soil S $\times$ time interactions need to be evaluated. The important role of subsoil S in supplying crop needs for the element has been aptly demonstrated (2,10,12,22,31). Soil samples for available S assay should be collected assuming that sulphate acts as a mobile nutrient (9) and to the rooting depth of the crop to be grown.

Sample preparation can influence both chemical, and biological estimates of available S. Increases as great as ten-fold in the amount of sulphate extracted have been noted from drying at 100° C (45). Any physical manipulation of a soil sample can be expected to influence biological activity and biological estimates of S availability (3,4). Practicability, and the relatively minor effects of air-drying and grinding on other than the readily soluble sulphate, however, dictates that samples be air-dried and crushed.

SULPHUR SOIL TESTS

Numerous procedures have been proposed for evaluating the S status of soils. Included are extraction of soil sulphate with water, various salt solutions; the release of sulphate upon incubation; microbial growth; and growth and S uptake by plants (see reviews by Freney *et al.* (16); Ensminger and Freney (13); Spencer and Freney (39); Beaton *et al.* (5); and Reisenauer *et al.* (34)). Soil S extractants can be categorized on the basis of S removal into 3 groups: 1. those that remove readily soluble sulphate; 2. those that remove readily soluble plus portions of adsorbed sulphate; and 3. those that remove the readily soluble and adsorbed sulphate, plus portions of the organic S. With but few exceptions, all of the procedures proposed have been found to be significantly related to crop response to S fertilization or uptake of the element by the crop. The relationships reported from greenhouse experiments are, as would be expected, much closer than those from field trials (34). No one procedure has been consistently superior.

Not all extractants perform equally well on all soils. Extractants that remove readily soluble plus portions of adsorbed sulphate tend to be better-correlated than those removing only readily soluble sulphate. Adsorbed sulphate is removed by neutral and slightly alkaline extracts, and by extractants containing a strong replacing ion as phosphate. Excellent prediction of response in field trials has been reported for extractants that remove soluble, adsorbed and portions of the organic S. Included are the acetic acid-phosphate solutions employed by Cooper (12) and Hoeft *et al.* (20), and the pH 8.5

$NaHCO_3$ solution used by Bardsley and Kilmer (2). Acidic solutions should not be used with calcareous soils.

Other factors of importance in selecting a procedure include cost, analytical feasibility and time requirements. Thus biological evaluations using *Aspergillus* (27) and green plants (8,10,37) are of limited utility for routine assay. Incubation procedures paralleling those used in available N estimates have met with little success (3,18) as little sulphate is formed and its precise measurement is difficult.

Several techniques for quantitative determination of extracted S have been proposed (5). Of the various procedures, the turbidimetric determination of sulphate as $BaSO_4$ is the most widely used. The factors affecting this determination have been widely studied (see Beaton *et al.* (5)) and recently an automated, nephelometric procedure has been adapted for sulphate measurement (7). Other procedures found applicable to the determinations of the small amounts of S extracted include the reduction procedure of Johnson and Nishita (21), X-ray fluorescence (35), and procedures measuring Ba following $BaSO_4$ precipitations (17).

INTERPRETATION OF S SOIL TESTS

The soil-plant relations of S both differ from, and resemble, those of the other nutrient elements. Sulphur, like N, is a constituent of protein and has an important organic component in soils. Plant requirements for S are similar to those for P. Sulphur differs from N and P in that its cycling through the biosphere is relatively much more rapid, in that its reserve supplies in both soils and plants are meagre and require continuous replenishment, and in that the intensity of its reactions with soils vary over greater limits. It is these differences that complicate the interpretation of available soil S assays.

The atmosphere is a principal source of S to most soils. Annual additions in the precipitation of as much as 234 and as little as 2 kg S/ha have been reported (14,23,36). In addition considerable amounts of S are absorbed by both plants and soils as SO_2 (1,41), and reach soil and plant surfaces from sedimentation and scavenging of particulate sulphate (24). The annual atmospheric contributions of S to the land areas of the world have been estimated to be in excess of 100 million tons (C. C. Delwiche, pers. comm.) (23,24). In contrast annual releases from weathering of rocks approximate only 2 million tons (C. C. Delwiche, pers. comm.). These large additions of S would go toward meeting crop needs if they were uniformly distributed. Unfortunately this is not the case. Localization of the principal sources of atmospheric S combined with the relatively short residence times of H_2S, SO_2 and sulphate particles in the lower atmosphere—approximately one week (24)—results in highly variable rates of additions. As a result the S supplies for many areas are inadequate and for other areas they are excessive.

Reserve supplies of S in soils and plants, relative to crop requirements for the element, are lower than those of other nutrient elements. Although sulphate adsorption imparts a degree of buffering in acid, fine-textured soils, sulphate is quite readily leached from most agricultural soils (19). Likewise S accumulation and reuse by plants is limited (8). As a result S deficiencies develop within a relatively short period after the immediate supply is exhausted.

In evaluating the likelihood of S responses in an area the magnitude and timing of external additions should be considered first. Those of primary significance are from the atmosphere, irrigation waters, and other agricultural chemicals. Crop needs for S are met at relatively low levels. Annual additions of 10 kg S/ha in the precipitation are generally adequate to supply crop needs (2,15,22,26,42,44). Irrigation waters containing more than 4 ppm of the element will supply the needs of irrigated crops (28). Recommended rates of S fertilization are commonly between 10 and 30 kg/ha (6) and as little as 5 kg/ha may be adequate (22). Any small addition of the element may produce significant effects.

Table 5-5 give interpretive values reported from field trials for different soil S extractants. The values are from different soils and crops, from different sampling procedures, and from different statistical interpretations. In spite of these differences the values agree surprisingly well and indicate a high

TABLE 5-5
Interpretive guides for extractable soil S*

Extractant	Critical level of extractable S**	Reference
	ppm	
Readily soluble sulphate:		
Water (Legume-grass)	3	(43)
0.1 *M* $CaCl_2$ (Legume-grass)	3	(43)
0.01 *M* $CaCl_2$ (Pastures)	6	(12)
Readily soluble plus portions of adsorbed sulphate:		
M NH_4OAc	6–7	(29)
KH_2PO_4-500 ppm P		
(Pasture)	8	(12)
(Pasture)	4	(38)
$Ca(H_2PO_4)_2$-500 ppm P		
(Alfalfa)	10	(15)
(Corn)	8	(15)
NaOAc-HOAc (Morgan's solution)		
(Alfalfa)	12	(18)
0.5 *M* $NaHCO_3$		
(Cotton)	10	(25)
Readily soluble, adsorbed, and portions of organic sulphate:		
$Ca(H_2PO_4)_2$-2 *M* HOAc		
(Alfalfa)	10	(20)
NaH_2PO_4-2 *M* HOAc		
(Pasture)	10	(12)

* Data from field-trials only.
** No response likely above this level.

level of dependability of these tests. The choice of an extractant is in part dependent upon the characteristics of the soils encountered. Readily soluble sulphate gives useful estimates in semi-arid areas for near neutral, low organic matter soils. In contrast, the acid-phosphate solutions are favoured in areas of acid, variable organic content soils. These correlations and the utility of the tests can be improved by better sampling procedures and by including in their interpretation quantitative expressions for the effects of the other factors influencing the supply, absorption and utilization of the element by the crop.

REFERENCES

(1) Alway, F. J., Marsh, A. W. and Methley, W. J. (1937). *Proc. Soil Sci. Soc. Am.*, **2**: 229.
(2) Bardsley, C. D. and Kilmer, V. J. (1963). *Proc. Soil Sci. Soc. Am.*, **27**: 197.
(3) Barrow, N. J. (1961). *Aust. J. Agric. Res.*, **12**: 306.
(4) Barrow, N. J. (1969). *Soil Sci.*, **108**: 193.
(5) Beaton, J. D., Burns, G. R. and Platou, J. (1968). The Sulphur Institute, Washington D.C., Tech. Bull. 14.
(6) Beaton, J. D. and Fox, R. L. (1971). *Fertilizer Technology and Use*. 2nd edn, Soil Sci. Soc. Amer., Madison, Wisc., p. 335.
(7) Bettany, J. R. and Halstead, E. H. (1972). *Can. J. Soil Sci.*, **52**: 127.
(8) Bouma, D. and Dowling, E. J. (1962). *Aust. J. Agric. Res.*, **13**: 791.
(9) Bray, R. H. (1954). *Soil Sci.*, **78**: 9.
(10) Cairns, R. R. and Richer, A. C. (1960). *Can. J. Soil Sci.*, **40**: 246.
(11) Coleman, R. (1966). *Soil Sci.*, **101**: 230.
(12) Cooper, M. (1968). *Trans. VII Int. Congr. Soil Sci.*, **2**: 263.
(13) Ensminger, L. E. and Freney, J. R. (1966). *Soil Sci.*, **101**: 283.
(14) Eriksson, E. (1960). *Tellus*, **12**: 63.
(15) Fox, R. L., Olson, R. A. and Rhoades, H. F. (1964). *Proc. Soil Sci. Soc. Am.*, **28**: 243.
(16) Freney, J. R., Barrow, N. J. and Spencer, K. (1962). *Pl. Soil*, **17**: 295.
(17) Galindo, G. G., Appelt, H. and Schascha, E. B. (1969). *Proc. Soil Sci. Soc. Am.*, **33**: 974.
(18) Harward, M. E., Chao, T. T. and Fang, S. C. (1962). *Agron. J.*, **54**: 101.
(19) Harward, M. E. and Reisenauer, H. M. (1966). *Soil Sci.*, **101**: 326.
(20) Hoeft, R. G., Walsh, L. M. and Keeney, D. R. (1973). *Proc. Soil Sci. Soc. Am.*, **37**: 401.
(21) Johnson, C. M. and Nishita, H. (1952). *Anal. Chem.*, **24**: 736.
(22) Jordan, H. V., Pearson, R. W. and Bardsley, C. E. (1960). *Trans. VII Int. Congr. Soil Sci.*, **3**: 563.
(23) Junge, C. E. (1960). *J. Geophys. Res.*, **65**: 227.
(24) Kellogg, W. W., Cadle, R. D., Allen, E. R., Lazrus, A. L. and Martell, E. A. (1972). *Science*, **175**: 587.
(25) Kilmer, V. J. and Nearpass, D. C. (1960). *Proc. Soil Sci. Soc. Am.*, **24**: 337.
(26) Lutz, J. A. Jr (1957). Va. Agr. Exp. Sta. Bull. No. 482.
(27) Malavolta, E. and Galli, F. (1951). *Ann. Esc. Agr. Queiroz*, **8**: 171.
(28) Martin, W. E. (1958). *Calif. Agr.*, **12**(11): 10.
(29) McClung, A. C., De Freitas, L. M. M. and Lott, W. L. (1959). *Proc. Soil Sci. Soc. Am.*, **23**: 221.
(30) McIntyre, G. A. (1967). *J. Aust. Inst. Agric. Sci.*, **33**: 309.
(31) Pawluk, S. and Bentley, C. F. (1964). *Can. J. Soil Sci.*, **44**: 261.
(32) Peck, T. R. and Melsted, S. W. (1973). *Soil Testing and Plant Analysis*. Revised edn, Soil Sci. Soc. Amer., Inc., Madison, Wisc., p. 67.
(33) Reed, J. F. (1953). *Better Crops with Plant Food*, **37**(8): 13.
(34) Reisenauer, H. M., Walsh, L. M. and Hoeft, R. G. (1973). *Soil Testing and Plant Analysis*, Revised edn, Soil Sci. Soc. Amer., Inc., Madison, Wisc., p. 173.
(35) Roberts, S. and Koehler, F. E. (1968). *Soil Sci.*, **106**: 164.

(36) Rossiter, R. C. (1952). *Aust. J. Agric. Res.*, **3**: 7.
(37) Sanford, J. O. and Lancaster, J. D. (1962). *Proc. Soil Sci. Soc. Am.*, **26**: 63.
(38) Spencer, K., Bouma, D. and Moye, D. V. (1969). *Aust. J. Exp. Agric. Anim. Husb.*, **9**: 320.
(39) Spencer, K. and Freney, J. R. (1960). *Aust. J. Agric. Res.*, **11**: 948.
(40) Stevenson, I. L. (1956). *Pl. Soil*, **8**: 170.
(41) Swinbank, W. C. (1968). *Quart. J. Roy. Meteorol. Soc.*, **94**: 460.
(42) Walker, D. R. (1969). *Can. J. Soil Sci.*, **49**: 409.
(43) Walker, D. R. and Doornenbal, G. (1972). *Can. J. Soil Sci.*, **52**: 261.
(44) Whitehead, D. C. (1964). *Soils Fertil.*, **27**: 1.
(45) Williams, C. H. (1967). *Pl. Soil*, **26**: 205.
(46) Williams, C. H. and Steinbergs, A. (1964). *Pl. Soil*, **21**: 50.

Evaluation of Soil and Plant Analysis to Determine Sulphur Status of New Zealand Soils

W. M. H. SAUNDERS and D. M. COOPER

Farming in New Zealand is predominantly pastoral, and development of soil testing has been mainly concerned with the determination of fertilizer requirements of soils under pasture. There were early reports of sulphur deficiency affecting pastures and lucerne in New Zealand (8,30,31,32,33) but widespread use of superphosphate obscured the need for sulphur. In the early 1950s many field experiments were laid down to determine sulphur requirements for the establishment of white clover-ryegrass pastures (*Trifolium repens*, *Lolium perenne*) and these provided samples for developing soil and plant diagnostic methods.

Soil analysis

Miss Cooper (7) screened many of the early sulphur soil test methods, and selected five for comparison with results from field experiments. There were: Bardsley and Lancaster, 0.03 *M* $NaH_2PO_4.2H_2O$ in 2 *M* acetic acid (1); Ensminger, 500 ppm P as KH_2PO_4 (11); Williams and Steinbergs, 0.01 *M* $CaCl_2$ (38); Williams and Steinbergs, Heat-soluble S extracted with 1 per cent NaCl (38); Kilmer and Nearpass 0.5 *M* $NaHCO_3$ at pH 8.5 (19).

The first of these differentiated very well between responsive and non-responsive sites at a figure of 10 ppm S in air-dry soil, and it has been used on a limited scale in the soil testing service of the New Zealand Ministry of Agriculture and Fisheries. The second and third were almost as satisfactory but the latter two methods were unsatisfactory under New Zealand conditions.

Sulphate retention

New Zealand soils differ widely in their capacity to retain added sulphate (14,29) (Table 5-6) and leaching of sulphate from soils of low sulphate retention is common (10,15,16,35,36). The original laboratory method (14), using leaching, was slow, and in later work (29) an equilibration method, similar to Barrow's (4), has been used. The agreement between the two methods was good ($r = 0.87$) (29). Sulphate and phosphate retention are closely related (Table 5-7), as previously noted (4), and this enables the more simple phosphate retention measurement (28) to be used as an index of sulphate retention.

TABLE 5-6
Ranges of sulphate and phosphate retention values for topsoils and subsoils of soil groups of New Zealand (29)

Soil group (25)	% S retention		% P retention	
	0-75 mm	300-450 mm	0-75 mm	300-450 mm
Brown-grey earths	0–10	0–15	10–25	10–30
Yellow-grey earths	0–10	0–15	10–35	15–40
High country yellow-brown earths	0–30	20–60	30–60	20–90
Yellow-brown earths	0–50	0–80	20–60	10–90
Podzols, southern	0–20	0–60	0–20	10–80
Podzols, northern	0–10	0–70	0–5	0–100
Yellow-brown sands	0–20	0–80	0–60	0–95
Yellow-brown pumice soils	0–45	0–90	10–80	0–95
Yellow-brown loams	10–70	60–95	60–100	80–100
Brown granular loams, southern	0–30	n.a.	25–75	30–90
Brown granular loams, northern	0–40	85	35–90	50–85
Red and brown loams	10–50	70–80	60–90	60–95
Gley soils	0–10	11	15–30	10–15
Recent soils, alluvial	0–10	0–50	10–40	5–60
Recent soils, volcanic	0–10	5–15	15–60	10–40

Initially it was believed that rapid losses of sulphate would occur on all soils of low sulphate retention, but subsequent work has shown that the effective rainfall must be taken into account and that there must be a sufficient net flow of water through the soil either vertically or horizontally (9,10,29).

Plant analysis

Metson (22) has reviewed the world and New Zealand literature on this means of diagnosing soil sulphur status. Estimates of 'critical levels' made for New Zealand conditions (20,21) have shown that total sulphur concentration in pasture was influenced by the species present, their relative proportions, the plant part sampled, the stage of growth and interactions with the nitrogen concentration in the plant. A standardized sample, taken at grazing height from plants at a young stage of growth, is used and the critical level set for clovers is 0.25 to 0.30 per cent S in dry matter (with an overall average of 0.28 per cent S) and for grasses, 0.25 to 0.30 per cent S. The N:S ratio should also be used and its critical value is in the range 17 to 19 for white clover and 15 to 17 for ryegrass.

Recent work

In 1967 a project was commenced to determine the sulphur maintenance requirements for soils under pasture which had been intensively farmed

TABLE 5-7
Levels of sulphate retention in topsoils and subsoils corresponding to levels of phosphate retention (29)

% phosphate retention		% sulphate retention	
		Topsoil (0–75 mm)	Subsoil (300–450 mm)
Very high	86–100	30–60	30–60
High	61–85	0–30	30–50
Medium	31–60	0–30	10–30
Low	0–30	0–30	0–20

and regularly topdressed with superphosphate for many years. Samples from these trials are being used to evaluate soil and plant analysis as means of assessing sulphur status and fertilizer requirements. The plant analytical work is not yet complete and the remainder of this paper is a preliminary report on the evaluation of soil analysis. The field work of the project was being completed at the time of writing, so results and conclusions are tentative.

EXPERIMENTAL METHODS

Soils

Nine soil sets (24) were used (Table 5-8) covering a range of sulphate retention from high in the Waihou soil to low in the Hurunui soil. On each soil set, field experiments were laid down in eight paddocks which had a range of previous fertilizer usage from low to very high.

Field experiments

An essential feature of the field experiments was that pasture management was almost wholly in the hands of the individual farmer. This meant that the degree of response to added sulphur and hence the correlations with the laboratory soil test values, would be influenced not only by the level of available sulphur in the soil but also by farm to farm variation of plant factors, such as pasture species, grazing management and pressure, climatic factors, and by the natural return of urine and dung. However, these are the conditions in which the soil tests are used.

Responses to sulphur were measured by mowing cuts in the flush of spring growth (October to November) and in autumn when the soils had re-wet after summer drought (April to June). The plots were soil sampled (0-75 and 0-150) mm and top-dressed with rates of gypsum in August and February. Stock were excluded from the trial areas only for a short time immediately after top-dressing and for the period over which pasture growth was measured. For the rest of the year the plots were managed and grazed with the surrounding paddock.

Laboratory methods

Four methods for extraction of soil sulphur were used: 0.01 *M* $CaCl_2$, (38)

TABLE 5-8
Properties of soils used in field experiments (0-75 mm)

Soil set of type (23, 26)	Classification (25)	S retn %	P retn %	pH	Superphosphate usage (kg/ha/year)	Initial S soil test (ppm S)**	No. of observations	
							Total	Responses
Waihou sandy loam	Central yellow-brown loam	37–57	92–95	5.3–5.7	125–1000	22–50	54	6
Katikati sandy loam	Central yellow-brown loam	n.a.*	88–95	4.9–5.7	250–500	19–29	57	3
Tangitiki sandy loam	Northern yellow-brown sand†	14–40	65–92	5.5–5.9	125–600	20-58	34	1
Taupo sandy silt	Yellow-brown pumice soil	16–20	61–74	4.8–5.4	350–500	15–27	92	10
Oruanui sand	Yellow-brown pumice soil	0–18	43–64	5.4–6.0	250–750	7–32	52	21
Hauraki clay	Gley soil	n.a.	18–29	5.2–5.9	125–350	14–46	28	5
Matapiro silt loam	Central yellow-grey earth	0–3	14–27	5.4–6.4	180–300	14–24	57	2
Opuha silt loam	Southern yellow-grey earth	n.a.	13–24	5.5–6.7	125–250	9–16	48	17
Hurunui hill soil (stony loam)	Central yellow-brown earth, steepland	2–12	9–34	n.a.	0–180	4–24	33	22

* Not available.
** Phosphate in 0.35 *M* acetic acid.
† G. E. Orbell, pers. comm.

500 ppm P as KH_2PO_4, (11) 0.03 *M* Na $H_2PO_4.2H_2O$ in 2 *M* acetic acid (pH 2.0) (1) and 0.03 *M* $NaH_2PO_4.2H_2O$ in 0.35 *M* acetic acid (pH 3.0).

The 0.01 *M* Ca Cl_2 extraction was discontinued after the first year because it was underestimating the sulphur status of the more highly sulphate retentive soils, probably because adsorbed suphate was not extracted (2,13). Phosphate in 2 *M* acetic acid (pH 2) was similarly not fully estimating the sulphur status of the high sulphate retentive soils and it was modified (method 4). The soil: solution ratio of 1:5 and the concentration of sodium phosphate at 0.3 *M* were retained but the time of extraction was increased to one hour and the pH increased. Extraction of sulphur from the highly retentive soils increased but there was no change for soils of low sulphate retention. Calcium phosphate solutions (3,12,37) showed no advantage over the potassium and sodium phosphate solutions even on soils of high organic matter content (4 to 12 per cent C).

Sulphur was initially determined turbidimetrically (6) after oxidation of the extracts with H_2O_2. Later a modified Johnson and Nishita procedure was used (5,18,27).

RESULTS AND DISCUSSION

Most of the 72 field experiments were carried on for four years and at each site at least six pasture yield measurements were made over the period. In evaluating the usefulness of the chemical soil tests, each measurement of pasture growth was regarded as an independent observation.

On six of the soils, Waihou, Katikati, Tangitiki, Taupo, Hauraki and Matapiro, there were very few responses to added sulphate (Table 5-8). This result, indicating a large residual effect of previous fertilizer, was not unexpected on the soils of high sulphate retention, Waihou, Katikati, Tangitiki and Taupo. However, both the Matapiro and Hauraki soils have low sulphate retention of the same order as that of the Oruanui sand (Table 5-8). On the latter soil (15) leaching of sulphate over the winter period is a serious problem and in the present series of experiments, sulphur deficiency developed at most sites on this soil within two years. The rainfall for the three soils is of a similar order (850 mm on Matapiro, 1,200 mm on Hauraki and Oruanui) with a very dry spell generally during January and February. In contrast to Oruanui sand, the Hauraki and Matapiro soils have a heavier texture (Table 5-8) and their drainage is slower in winter. In addition, there may be upward movement of water during the period of summer drought in these profiles (Saunders and Hogg, unpublished data). Both these factors would reduce leaching of sulphate.

Relation between soil test values and response to sulphur

This discussion is limited to the results from the 0-75 mm soil samples. Present results indicate no advantage in using 0-150 mm samples. Correlation coefficients between soil test values and relative yields were determined for the Oruanui and Opuha soils only because of lack of response on the remaining soils, except Hurunui. Responses on Hurunui soil were measured by visual pointing which could not be expressed quantitatively as relative yields.

The coefficients for the Oruanui and Opuha soils were generally poor varying between years and seasons from 0.2 to 0.8 and were rarely significant. This lack of correlation is probably largely because the plant and climatic factors strongly influenced the degree of response to sulphur under the system of individual management of the trial sites. However, the relationship between soil test values and field responses may be evaluated more simply on the basis of per cent probability of correct prediction of response or no-response at a given soil test value. The number of observations of response or no-response occurring at each soil test value was recorded and then the following ranges were selected by inspection with the object of giving the best differentiation between responsive and non-responsive sites:

1. very low 0-5 ppm S in air-dried soil;
2. low 6-9 ,, ,, ,, ,, ;
3. medium 10-15 ,, ,, ,, ,, ;
4. high 15 ,, ,, ,, ,, .

The per cent of the total number of observations within each range, that were responsive to added sulphur, is shown in Table 5-9A. The incidence of response decreases as the soil test values rise, but there is no clear-cut critical level. In general agreement with the early work (7) the incidence of response is small at values of 10 ppm S and greater, by all three methods, and decreases at values greater than 15 ppm S. In the very low range (0-5 ppm S), phosphate in 0.35 *M* acetic acid gives a very good prediction of response but the number of observations is small. Ensminger's reagent is less useful and phosphate in 2 *M* acetic acid is unsatisfactory. Over the low range (6-9 ppm S) none of the methods gives any differentiation between responsive and non-responsive sites.

Overall, phosphate in 0.35 *M* acetic acid is a little more useful than Ensminger's reagent, particularly for responsive sites. The previously recommended, phosphate in 2 *M* acetic acid, is useful with soils of low sulphate retention (Table 5-9B) but in the highly retentive soils (Table 5-9C) it does not extract enough adsorbed sulphate to indicate fully the sulphur status. The early work (7) was done mainly with soils of low sulphate retention and probably many extractants such as 0.01 *M* $CaCl_2$ and water would similarly be useful on such soils but would fail where sulphate retention is high and adsorbed sulphate is the dominant form of sulphate present.

Adsorbed plus water-soluble sulphate appear as the main sources of sulphur for pastures under New Zealand conditions. This is to some extent surprising in this situation, because there is almost continuous cycling of sulphur through the soil, plant and grazing animal, and it might be expected that the best estimate of sulphur status would have included a larger proportion of the soil organic sulphur than the small amount extracted by the phosphate reagents.

Adsorbed plus water-soluble sulphate is, however, normally only a small part of the total soil sulphur and it may be regarded as a pool (34) which gains and loses sulphur by processes such as breakdown of plant and animal residues, mineralization of organic forms of sulphur, uptake by pasture, leaching, and immobilization in soil organic matter. The amount of sulphate in this pool of immediately available sulphur is probably an index of the total amount of sulphur present in an 'active' cycle in the pasture ecosystem. Where losses from this 'active' cycle are large, by leaching or by immobilization in soil organic matter, such as in the Oruanui soil (15,17) then fertilizer sulphur is required to maintain an adequate level of sulphur in the whole of the cycle. Where the two latter processes are at a minimum, in long established pastures on soils of high sulphate retention (Waihou, Katikati, Tangitiki and Taupo) or on soils where there is little net movement of water for leaching (Matapiro and Hauraki), little or no fertilizer sulphur may be required for several years.

TABLE 5-9

Total number of observations and per cent responsive to fertilizer sulphur within four ranges of laboratory soil test values

Soil test values (ppm S in air-dry soil)	P in 0.35 *M* acetic acid		Ensminger reagent		P in 2 *M* acetic acid	
	Observations		Observations		Observations	
	Total	% Response	Total	% Response	Total	% Response
A. All soils						
0–5	12	92	12	67	52	40
6–9	75	52	78	51	161	26
10–15	124	18	112	20	180	11
15	242	6	253	7	62	6
B. Soils of low sulphate retention						
0–5	11	91	12	67	20	80
6–9	68	56	75	51	85	39
10–15	85	19	87	22	84	19
15	53	6	44	5	29	7
C. Soils of high sulphate retention						
0–5	1	100	0	—	32	16
6–9	7	14	3	67	76	12
10–15	39	15	25	12	96	4
15	189	6	209	7	33	6

While some improvement of the estimate of sulphur status might be expected by including an estimate of elemental sulphur or sulphate at depth in the soil profile, a real improvement will probably only be achieved with much better knowledge of the forms of sulphur present in the 'active' cycle. However, the major limitation to the usefulness of laboratory soil testing in predicting the degree of response and fertilizer requirements of soils under pasture is the lack of quantitative data on the effects of climatic and plant factors on the capacity of a pasture to respond to fertilizer.

Acknowledgements. The authors wish to thank the many scientists and technicians of the Research Division of the New Zealand Ministry of Agriculture and Fisheries, who carried out field and laboratory work and statistical analysis for this project.

REFERENCES

(1) Bardsley, C. E. and Lancaster, J. D. (1960). *Proc. Soil Sci. Soc. Am.*, **24**: 265.
(2) Barrow, N. J. (1960). *Aust. J. Agric. Res.*, **12**: 306.
(3) Barrow, N. J. (1967). *Soil Sci.*, **104**: 242.
(4) Barrow, N. J. (1967). *Soil Sci.*, **104**: 342.
(5) Beaton, J. D., Burns, G. R. and Platou, J. (1968). Sulphur Inst. Tech. Bull. 14: 42.
(6) Chesnin, L. and Yien, C. H. (1950). *Proc. Soil Sci. Soc. Am.*, **15**: 149.
(7) Cooper, M. (1968). *Trans. IX Int. Congr. Soil Sci.*, Adelaide, **2**: 263.
(8) Doak, B. W. (1929). *N.Z. J. Sci. Technol.*, **11**: 25.
(9) During, C. (1970). *N.Z. Soil News*, **18**: 72.
(10) During, C. and Martin, D. J. (1968). *N.Z. J. Agric. Res.*, **11**: 665.
(11) Ensminger, L. E. (1954). *Proc. Soil Sci. Soc. Am.*, **18**: 259.
(12) Fox, R. L., Olson, R. A. and Rhoades, H. F. (1964). *Proc. Soil Sci. Soc. Am.*, **28**: 243.
(13) Hasan, S. M., Fox, R. L. and Boyd, C. C. (1970). *Proc. Soil Sci. Soc. Am.*, **34**: 897.
(14) Hogg, D. E. (1965). *N.Z. Grassld Assoc. Proc.*, **27**: 123.
(15) Hogg, D. E. and Toxopeus, M. R. J. (1966). *N.Z. J. Agric. Res.*, **9**: 93.
(16) Hupkens van der Elst, F. C. C. (1968). *Trans. IX Int. Congr. Soil Sci.*, Adelaide, **3**: 407.
(17) Jackman, R. H. (1964). *N.Z. J. Agric. Res.*, **7**: 445.
(18) Johnson, C. M. and Nishita, H. (1952). *Anal. Chem.*, **24**: 736.
(19) Kilmer, V. J. and Nearpass, D. C. (1960). *Proc. Soil Sci. Soc. Am.*, **24**: 337.
(20) McNaught, K. J. (1970). *Proc. XI Int. Grassld Congr.*, p. 334.
(21) McNaught, K. J. and Chrisstoffels, P. J. E. (1961). *N.Z. J. Agric. Res.*, **11**: 177.
(22) Metson, A. J. (1973). Sulphur Inst. Tech. Bull. 20.
(23) N.Z. Soil Bureau (1954). Bull. (n.s.) 5.
(24) N.Z. Soil Bureau (1962). Bull. 25, p. 143.
(25) N.Z. Soil Bureau (1968). Bull. 26.
(26) N.Z. Soil Bureau (1968). Bull. 27.
(27) Rees, T. D., Gyllenspetz, A. B. and Docherty, A. C. (1971). *Analyst*, **96**: 201.
(28) Saunders, W. M. H. (1965). *N.Z. J. Agric. Res.*, **8**: 30.
(29) Saunders, W. M. H. and Hogg, D. E. (1971). Ruakura Tech. Report No. 1.
(30) Tennant, R. B. (1927). *N.Z. J. Agric.*, **34**: 329.
(31) Tennant, R. B. and Duff, A. S. (1929). *N.Z. J. Agric.*, **39**: 320.
(32) Tennant, R. B. and Hume, A. A. (1928). *N.Z. J. Agric.*, **36**: 173 and 245.
(33) Tennant, R. B. and Hume, A. A. (1928). *N.Z. J. Agric.*, **37**: 38.
(34) Till, A. R. and May, P. F. (1970). *Aust. J. Agric. Res.*, **21**: 253.
(35) Toxopeus, M. R. J. (1970). *Proc. XI Inst. Grassld Congr.*, p. 345.
(36) Toxopeus, M. R. J. and Gordon, R. B. (1971). *N.Z. Grassld Assoc. Proc.*, **33**: 35.
(37) Williams, C. H. (1968). *Aust. J. Soil Res.*, **6**: 131.
(38) Williams, C. H. and Steinbergs, A. (1959). *Aust. J. Agric. Res.*, **10**: 340.

Evaluation of Plant and Soil Sulphur Tests in Australia

C. S. ANDREW

This is a summary of presentations on the evaluation of tests in Australia made by the six state representatives at the Symposium.

Space limitations precluded their presentation in full and the reader is referred to the *Handbook on Sulphur in Australian Agriculture* for most of the basic information. Acknowledgement is made of the various contributors: Queensland—C. S. Andrew, B. J. Crack, and G. E. Rayment; New South Wales—J. S. Glendinning, R. G. Weir and M. H. Walker; Victoria—D. R. Rooney, J. Hosking, and G. P. Briner; Tasmania—W. G. Martinick; Western Australia—P. G. Southern and J. W. Bowden; South Australia—D. C. Lewis and A. L. Clarke.

Standards for the derivation of 'critical values' were obtained by two methods. Firstly, short-term experiments done using numerous rates of application to a particular soil/s and the observed plant growth response were related to some soil/plant chemical factor. The resultant 'critical concentration' is accepted as tentative. Secondly, standards may be calculated from collections of data, gathered over a longer time period, and wider variety of soil types, where a minimum number of treatments (+ and —) have been used, and the associated plant response data have been recorded.

In the presentations at the Symposium, most of the standards were obtained by the second method. In some instances a range of values was used in preference to a single critical value.

SOIL CHEMICAL METHODS

Total sulphur

Although some attempts have been made to use total S as a soil test, the majority of investigations have given very poor correlations between soil values and plant growth response. In Queensland responsive soils were very low in total sulphur. The mean for 37 responsive soils, covering several crops, was 0.013 per cent and for non-responsive soils was 0.056 per cent. These means had wide limits and may not be very useful except in soils of similar character. For example, at the Queensland Wheat Research Institute, responsive soils varied between 0.014 and 0.058 per cent total S and the non-responsive soils between 0.015 and 0.06 per cent S. Workers in Western Australia, South Australia and New South Wales have shown very poor correlations between total soil S and plant growth response.

Some New South Wales workers believed better correlations would be obtained within a soil type or Great Soil Group, whilst in Victoria emphasis was placed on soil texture.

'Available' soil sulphur

A number of chemical extractants, e.g., $Ca(H_2PO_4)_2$, $CaCl_2$, $KHCO_3$, KCl, HCl have been used but use of calcium dihydrogen phosphate has received the most attention. The surface soil (0 to 10-15 cm) has been the usual portion of the profile to be sampled. Collective data from Queensland, and covering many unrelated soils and experiments, show that all responsive soils had SO_4-S concentrations of less than 13 ppm. A more detailed and related batch of heavy textured soils from the Queensland Wheat Research Institute provided a threshold sulphur value of 10 ppm SO_4-S. In Queensland there is evidence of differential influences by the plant species grown.

The few correlative studies made in New South Wales have not been promising for diagnostic purposes. Likewise, the Victorian workers do not consider the calcium phosphate, unmodified, an effective extractant. Tasmanian work on the use of phosphate extractable SO_4 as a diagnostic tool was not encouraging. Serious complications arose from leaching water, depth of soil sampled, relative accumulation of SO_4 in various portions of the profile, and from the type of pasture or crop. In Western Australia there is a serious leaching problem in many areas and most of the soils have a very low SO_4 adsorptive capacity. These factors combine to give a low correlation between displaced SO_4 and crop response. Research in South Australia has not provided any good correlations for the SO_4 test. Some Queensland workers have measured 'phosphate extractable S' for different soil horizons and found that the extractable S increases with soil depth. Knowledge of this type has helped to interpret experimental results in several states but no attempts have been made to use the information for diagnostic purposes.

Interference from organic sulphur in these extracts has been noted and workers have treated the extract with charcoal to remove it. Victorian workers are developing this technique, have tested 98 soils and arrived at a critical concentration of 5.4 ppm SO_4-S in the clarified extract.

There have been few tests with $CaCl_2$, and KCl as extractants and no good correlations with plant response have been obtained with any of them.

Western Australian researchers have used a stronger acid extractant, 0.01 *M* HCl. This extracts the solution and adsorbed phases and portion of the organic and insoluble inorganic components. Tests on a number of soils indicate that soils with concentrations of up to 100 ppm are very responsive soils, 200 to 350 ppm moderate to low response soils, and 200 to 600 ppm no response soils. A diagnostic value of 200 ppm in the HCl extract is accepted, however, the overlap of the two latter groups causes some concern, and further research will continue into this method.

The consensus of opinion was that soil tests were not very efficient at predicting the degree of plant response to added sulphur. Serious complicating factors include soil types, soil water relations prior to or at the time of sampling, the depth of sampling, plant species and analytical techniques. In general sulphur deficiencies occur in two broad groups of soils; soils of very light texture, well drained, with low organic matter content; and soils of medium to heavy texture with relatively high adsorptive capacity for SO_4, usually from low rainfall areas. Such soil differences may demand different methods of sufficiency assessments and/or different critical concentrations. In view of the incidence of leaching rains, field water relations at the time of sampling need to be taken into account. Samples should either be taken at the close of, or immediately prior to the commencement of the growing season. In attempts to improve correlations some research workers have standardized conditions of sampling and soil storage. Thus the Victorian, New South Wales and West Australian workers are attempting to counter the effects of longer soil storage times, storage temperatures and soil moisture content during storage on the extractable SO_4-S determined. Good results have been obtained with immediate drying at 40°C before storage.

'Available' soil phosphorus

New South Wales investigators have attempted to make use of 'available phosphate' determinations as an aid in making recommendations for sulphur applications, especially where plant growth response to superphosphate is obtained. This was based on the knowledge that characteristics of PO_4 and SO_4 adsorption on soil surfaces have some similarities. Using a modified Olsen method of extraction for 'available phosphate', recommendations are made on the basis of the severity of the phosphate deficiency. Recommended values were: soils with less than 35 ppm extractable P, use single superphosphate; soils with values of 35 to 45 ppm, use sulphur fortified single superphosphate (SF 25); and for soils with 45 ppm P or more use SF 45 or gypsum. With gypsum a biennial application of approximately 70 to 80 kg S/ha is recommended. The authors qualify the use of this system relative to soil types and a general knowledge of the areas concerned.

General

No common acceptable method of 'available' sulphur determinations arose from the collective data presented by the representatives from the six states. Nor could a common reference standard of the critical concentration be determined. This is not surprising when the extremes of soil type, the different environmental conditions (tropical to temperate), the large range of plant species involved and the varied management systems encountered are considered. Research workers need to develop correlations with a particular method of soil S diagnosis for a given situation and determine a range of critical values for the principal plant species grown there.

PLANT CHEMICAL METHODS

Total sulphur

Workers in Queensland determined tentative critical concentrations in some tropical and some temperate pasture legumes on a minimum number of soils in short-term experiments. The values vary for the different species ranging from 0.13 per cent for *Stylosanthes humilis* to 0.18 per cent for *Glycine wightii* in the tropical group, and from 0.16 to 0.21 per cent in the temperate group of legumes. For agronomic species, sugar cane, tobacco, etc., plant response to one level of applied S, usually under field conditions on several soils, has been related to S content and the minimum concentrations in the non-responsive range from the pooled data accepted as the tentative critical concentration.

In New South Wales total S analysis of plants has been used for pastures, maize, orchard samples, and some other crops. For *T. repens* and *T. subterraneum* the critical concentration was 0.26 per cent. Field data indicated that leaves of responsive *M. sativa* contained < 0.20 per cent S (range 0.10 to 0.19 per cent). *M. sativa* not responding to S addition had plant S concentrations of > 0.20 per cent (range 0.20 to 0.46 per cent), eleven trials. With pasture species, the 0.20 per cent level can be applied to most clovers. However, for lucerne, they classify plant S concentrations of < 0.2 per cent as deficient, 0.2 to 0.3 per cent marginally deficient, and > 0.3 per cent sufficient. In maize and sorghum a critical concentration of 0.12 per cent S is accepted, for brassicas 0.7 per cent, and 0.10 per cent for leaves of stone fruit trees. Sulphur deficiency has not been recorded in stone or citrus fruit trees. Victorian workers accept a working arbitrary S critical concentration of 0.20 per cent for most species. They suggest that species should be analysed separately and a value of 0.18 per cent has been recommended for *T. repens* and 0.2 per cent for grasses sampled in mid-spring. In tobacco, 0.25 per cent S in the leaf marks the point below which beneficial responses to applied S can be expected, whereas values in excess of about 0.3 per cent may be detrimental to plant quality. Citrus leaves commonly contain 0.2 to 0.3 per cent S, while deciduous fruit leaves usually contain 0.1 to 0.2 per cent. No attempt was made at a diagnostic index for these crops.

Western Australia has been primarily concerned with pasture species, and a critical concentration of 0.18 per cent is quoted for subclover. However, there are differences between cultivars, e.g., in one cultivar, deficiency occurred at 0.21 per cent and sufficiency at 0.23 per cent; in another cultivar, deficiency occurred at 0.13 per cent and sufficiency at 0.16 per cent S.

In South Australia a critical concentration in a mixed clover-grass system of 0.18 to 0.20 per cent is accepted.

Sulphate sulphur

Very few Australian workers have attempted to use SO_4-S in plants for diagnostic purposes, and as yet no attempt to determine an SO_4-S diagnostic index has been recorded.

Nitrogen:sulphur ratios

A critical value for this ratio is usually quoted at 15:1. In Queensland, values of 14 and 16 have been obtained as a diagnostic index for *M. sativa*. In Western Australian investigations, while no attempt was made to determine a diagnostic index, N:S values for subclover tops ranged between 14 and 29 in the absence of applied S and from 11 to 17 in the presence of applied S.

New South Wales workers found very wide N:S ratios (3.2:1 to 21:1) in lucerne leaf sample, but considered that a ratio of 14 to 16 was adequate for healthy plant growth. Ratios above this signify S deficiency.

Tasmanian results were based on N:S in the protein fraction, (N:S)p, and on total N:S ratios, (N:S)t. Sulphur deficiency prevailed when (N:S)t in the foliage equalled or was less than (N:S)p for the same plant and conditions. However, the (N:S)t ratio proved a more useful index than the (N:S)p. In ryegrass S deficient plants had (N:S)t values of approximately 37 whereas S sufficient plants had values of approximately 9 to 12. For *T. repens*, S sufficiency (N:S)t values were 14 to 16 or less. In *Brassica napus*, adequately supplied with nitrogen, (N:S)t in the foliage of 11 or more indicated sulphur deficiency.

Protein nitrogen:nitrate nitrogen ratios

Queensland workers have shown, in *S. humilis* and *M. lathyroides*, that protein N accounted for 70 to 91 per cent of the total N depending on the sulphur status. Increasing S supply increased the proportion of protein N. However, no attempt has been made to develop a diagnostic index. In Tasmania this ratio has been used for oats, ryegrass and *T. repens*. However, the ratio was not wholly reliable and the method is too laborious for routine diagnosis.

General

The consensus of opinion from all states was that the analysis of the plant foliage for total S was preferable to the analysis for fractions of S and/or nitrogen:sulphur ratios in their various forms, due mainly to the analytical difficulties associated with analyses involving fractionation.

Data on the conditions of plant sampling have been provided in the *Sulphur Handbook*. The majority of workers used the immediate preflowering stage, but some, e.g., Western Australia, suggested definite calendar timing and August has been used. A few workers suggested that sampling should be restricted to a specific six-week period in the growing season. Most of the data supplied related to whole plant tops; little attention was given to sampling particular plant parts. However, Western Australia did recognize the importance of developing indices for particular plant species, even cultivars.

RELATIONSHIP BETWEEN SOIL AND PLANT ANALYSIS

Victorian workers attempted to relate soil and plant analysis. They found generally good agreement between the soil and plant tests on deficient sites except where deeply rooted species such as lucerne, oats, phalaris and cocksfoot were grown. In these cases the usual 0-10 cm soil sampling depth was insufficient for estimating the sulphur requirement of these species.

The overall data presented at the Symposium indicated that both soil and plant diagnosis have merit. While there was some bias in favour of plant analysis it was considered that the approach to diagnosing sulphur status in the soil-plant-animal system should not be competitive. Ideally, the various methods of diagnosis should complement each other.

REFERENCE

Handbook on Sulphur in Australian Agriculture (1974), McLachlan, K. D. (ed.), CSIRO, Melbourne.

6 Fertilizer Technology

Types of Sulphur Fertilizers Made in Australia and Their Place in Agriculture

G. K. BELL

THE IMPORTANCE OF SULPHUR

Sulphur deficiencies in plants are occurring with increasing frequency in many important agricultural areas throughout the world. This is largely due to such factors as the increasing use of sulphur-free fertilizers, increasing crop and pasture requirements for sulphur resulting from increasingly higher yields; and, probably most importantly from the Australian viewpoint, the recognition of sulphur as a major plant nutrient. The importance of sulphur had been masked in the past by the common use of single superphosphate as the only fertilizer. This applies particularly to undisturbed pastures where the nutrients are continually taken from the top few inches of soil.

The history and uses of fertilizers in Australia is very largely that of superphosphate, at least up until the advent of 'high analysis' compound fertilizers and concentrated supers in the middle 1960s. This is no doubt responsible for the fact that the occurrence of sulphur deficiency in Australian soils was first reported as late as 1949 because until then the responses obtained by using superphosphate were credited to its phosphorus content, with no recognition of the sulphur fraction. It is now recognized that there are areas in Australia where a sulphur deficiency exists, and in some cases more sulphur than is contained in ordinary superphosphate must be applied to satisfactorily overcome this deficiency.

SOURCES OF SULPHUR AVAILABLE TO THE INDUSTRY

'Natural' sources include elemental sulphur, sulphur containing pyritic materials, naturally-occurring sulphates of calcium and potassium and mixed potassium/magnesium sulphates (langbeinite), sulphide ore of various minerals, sulphur containing crude oil and coal.

'Derived' forms include sulphuric acid itself and ammonium sulphate from scrubbing of coke oven gases, or from other chemical reactions.

A brief treatment of the way in which both elemental sulphur and ammonium sulphate are made available to the agricultural industry is presented.

While the majority of elemental sulphur was formerly mined from geological anticlinical structures called 'domes' by the Frasch Process, a rapidly increasing proportion is now recovered from 'sour' natural gas—particularly in France and Canada.

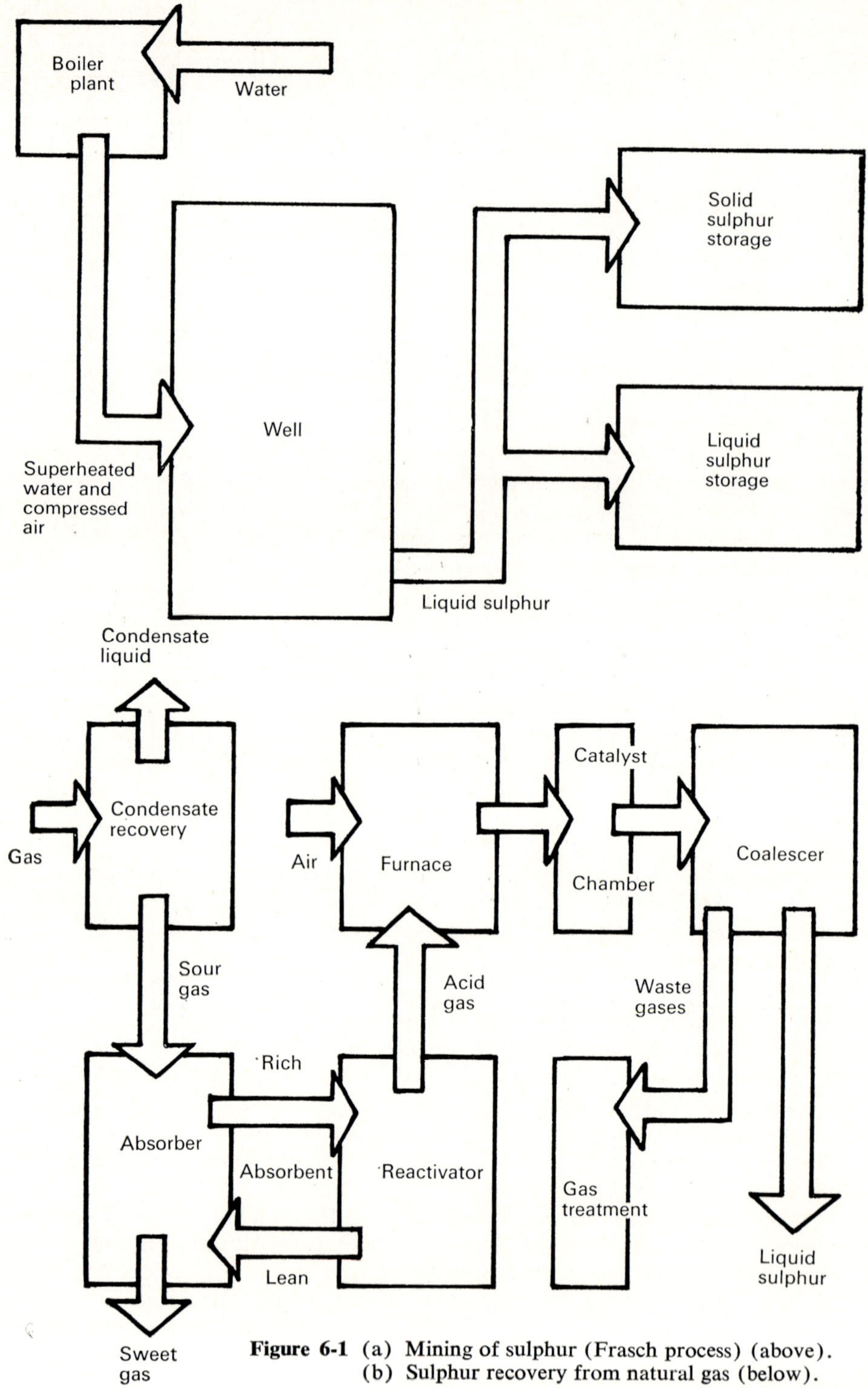

Figure 6-1 (a) Mining of sulphur (Frasch process) (above).
(b) Sulphur recovery from natural gas (below).

The Frasch mining process is remarkable in its simplicity. Superheated water is forced into an underground sulphur deposit. The water melts the sulphur, and the sulphur, more than 99 per cent pure, is raised to the surface.

Figure 6-1(a) shows the principal features of the Frasch Process (1).

Before it can be sold, sour natural gas must be sweetened; that is, the hydrogen sulphide must be removed. Therefore, as the gas flow at the well increases, sulphur production goes up automatically. Depending on the amount of hydrogen sulphide in the gas, the sulphur recovered is either a by-product, co-product or main product in the gas processing operation.

Sour natural gas is processed as shown in Figure 6-1(b) (1).

The process involves removing liquids from the gas, absorbing the hydrogen sulphide, stripping this out and burning with a catalyst after which the sulphur condenses to liquid form and is either stored in steam heated tanks or allowed to cool and solidify in outdoor vats.

Sulphate of ammonia is produced by 'scrubbing' of gases emanating from coke ovens operation, and is usually associated with an integrated steelworks such as the two large complexes in New South Wales. Ammonia is a constituent of this gas and following contact with sulphuric acid, ammonium sulphate is formed and purified by crystallization. Due to the presence of many other organic substances such as benzole, tar, pyridine, naphthalenes, etc., emanating from the coal, the total recovery plant can become quite complex, but the basic flow sheet is shown in Figure 6-2(a).

Another source of by-product sulphur is the gypsum produced during the reaction of sulphuric acid with phosphate rock in the manufacture of wet process phosphoric acid for later processing into fertilizers. Figure 6-2(b) gives the basic process.

This calcium sulphate can be applied as a source of sulphur and/or a soil conditioner or converted into ammonium sulphate or sulphuric acid. Large amounts have been used overseas in plasterboard and cement manufacture.

SULPHUR REQUIREMENTS BY PLANTS AND SOIL

Forms of sulphur used by plants

Plants usually absorb sulphur in the sulphate form. The soluble inorganic sulphates therefore supply this element in the form most quickly available to crops.

In many soils, however, the sulphate ion is not retained to any great extent. Soluble sulphates move with soil water and are readily leached from sandy soils under conditions of high rainfall. They are, however, adsorbed in the lower horizons of some soils.

Under certain conditions, elemental sulphur is oxidized to the sulphate form by soil micro-organisms. The rate of oxidation to plant-available sulphate depends on the extent of contact of the sulphur with the soil, temperature, moisture and aeration.

When non-water-soluble forms such as elemental sulphur are used, the particle size of the material is extremely important (2). To be oxidized rapidly in the soil, the material must be very fine (finer than 60 mesh) or, alternatively break down into very small particles after being applied. If adequate soil contact is made, elemental sulphur will provide ample plant nutrient

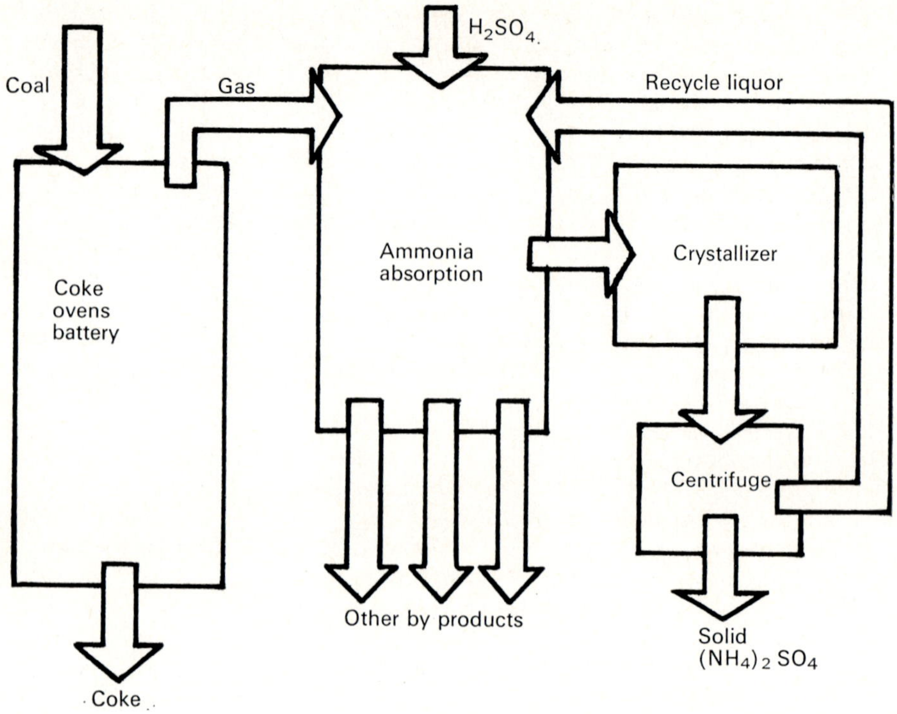

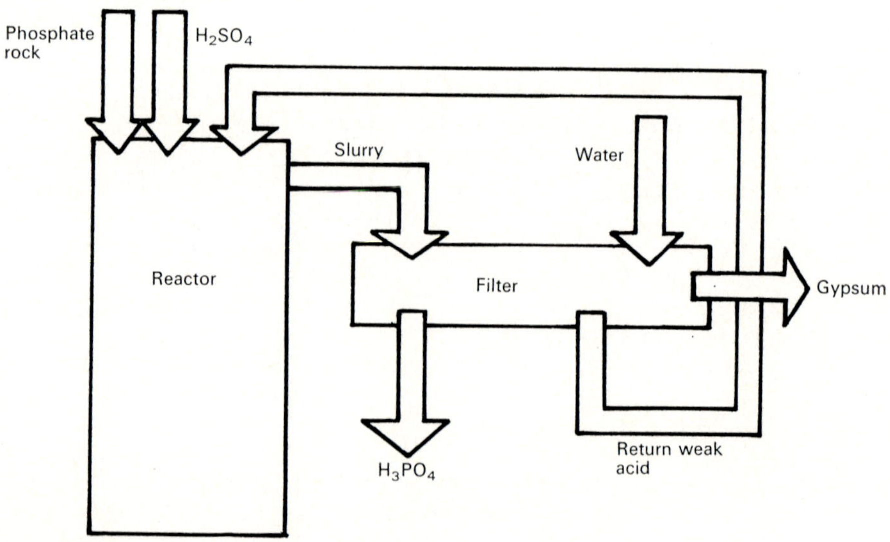

Figure 6-2 (a) Ammonium Sulphate from coke ovens gas (above).
(b) Phosphoric acid by wet process (below).

sulphur to crop and pasture plants. When sulphur is applied in liquid fertilizers, soil contact is usually extensive and conversion of reduced forms to sulphate follows quite rapidly. Optimum conditions for plant growth and for rapid sulphur oxidation are essentially the same.

TABLE 6-1
Sulphur-containing fertilizer materials in New South Wales

Material	Nitrogen (%)	Phosphorus (%)	Potassium (%)	Sulphur (%)	As elemental	As sulphate
'STRAIGHT' MATERIALS						
Sulphur	—	—	—	99.0	99.0	—
Ammonium sulphate	21.0	—	—	24.0	—	24.0
Potassium sulphate	—	—	48.0	16.0	—	16.0
Calcium sulphate	—	—	—	18.0	—	18.0
By-product gypsum	—	—	—	15.0	—	15.0
SUPERPHOSPHATES						
Superphosphate	—	9.5	—	12.0	—	12.0
Double superphosphate	—	17.5	—	5.0	—	5.0
Triple superphosphate	—	19.6	—	1.0	—	1.0
Super/Carb lime	—	4.6	—	6.0	—	6.0
Sulphur fortified '25'	—	7.7	—	27.0	18.0	9.0
Sulphur fortified '45'	—	5.9	—	45.0	37.0	8.0
Diammonium phosphate	16.0	21.0	—	1.0	—	1.0
COMPOUNDS						
12:22:0	12.4	22.3	—	1.5	—	1.5
15:13:0	15.0	13.1	—	10.0	—	10.0
20:5:0	19.6	4.9	—	19.0	—	19.0
18:8:0	17.5	8.1	—	16.0	—	16.0
6:12:23	6.6	12.0	23.2	2.0	—	2.0
11:15:9	11.0	14.7	9.0	5.0	—	5.0
12:8:15	12.0	7.9	14.9	9.0	—	9.0
14:4:15	13.6	4.4	14.7	12.0	—	12.0
16:7:6	15.6	7.0	6.3	12.0	—	12.0
MIXED FERTILIZERS						
5:8:4	5.0	7.5	4.2	14.0	—	14.0
10:4:6	10.0	3.9	6.2	17.0	—	17.0
9:4:5	9.2	4.2	4.6	17.0	—	17.0
10:5:0	10.5	4.8	—	18.0	—	18.0
Super/sulphate ammonia	10.5-5.2	4.6-6.9	—	18–15	—	18–15
Super/Muriate potash	—	4.7-7.8	24.9-8.3	6–2	—	6–2

TYPES OF FERTILIZER CONTAINING SULPHUR AND MANUFACTURING TECHNIQUES

The sulphur content of various 'straight' fertilizer materials is shown in Table 6-1. These include elemental sulphur, which is in excess of 99 per cent purity, ammonium sulphate with 24 per cent equivalent S, potassium sulphate with 16 per cent and calcium sulphate with 15 to 18 per cent, depending upon whether it is natural or by-product gypsum.

A large number of readily available 'manufactured' fertilizers also contain sulphur in varying proportions. This sulphur is incorporated in the fertilizer during the manufacturing process.

For convenience, these fertilizers can be divided into three groups: 1. superphosphates; 2. compounds and mixtures; and 3. sulphur fortified superphosphates.

Looking first of all at the superphosphates, ordinary or single superphosphate contains 12 per cent sulphur. This sulphur is present in the form of 'sulphate' sulphur, with the chemical form of calcium sulphate (anhydrite). The figure of 12 per cent refers to the sulphate sulphur content expressed as 'elemental S'.

This sulphate is introduced into the superphosphate when finely ground rock phosphate is acidulated with sulphuric acid to make the superphosphate. The amount of sulphuric acid introduced can be varied within limits but is virtually at a ratio such that the product superphosphate contains the equivalent of 12 per cent S in the form of calcium sulphate and 9.5 per cent P.

In an enriched superphosphate, made by acidulating phosphate rock with a mixture of phosphoric and sulphuric acids, naturally the sulphate content of the resultant enriched superphosphate is less than 12 per cent because the phosphoric acid used forms monocalcium phosphate only, rather than the mixture of monocalcium phosphate and calcium sulphate produced in the sulphuric acid reaction. Depending upon the ratio of the two acids used, fertilizers carrying varying grades of P can be produced, and as the concentration of P increases, the sulphate content decreases. An example of an enriched superphosphate is 'double superphosphate' which contains about 6 per cent S and 17 per cent P.

Thus, with triple superphosphate, which is made by treating phosphate rock with phosphoric acid alone, the sulphate content of the product is limited to the small amount arising from the sulphuric acid impurity in the wet process phosphoric acid used for acidulation.

Low analysis grades, such as 5:8:4 and 10:4:6, listed in Table 6-1 are examples of fertilizer 'mixtures'. These mixtures are usually made by mixing together 'straight' solid fertilizer materials such as superphosphate, sulphate of ammonia and/or potash salts and then granulating the resultant product. In some cases they inherit sulphur both from the superphosphate and from the sulphate radical from one of the salts, such as ammonium sulphate or potassium sulphate. A common fertilizer in this category is super/sulphate of ammonia in varying proportions, which is made either as indicated above, or by adding sulphate to fresh superphosphate 'ex den' and granulating direct. Figure 6-3(a) illustrates the different processes involved.

In the case of the more sophisticated ammonium phosphate based compounds, such as those listed in Table 6-1, sulphur may be introduced from

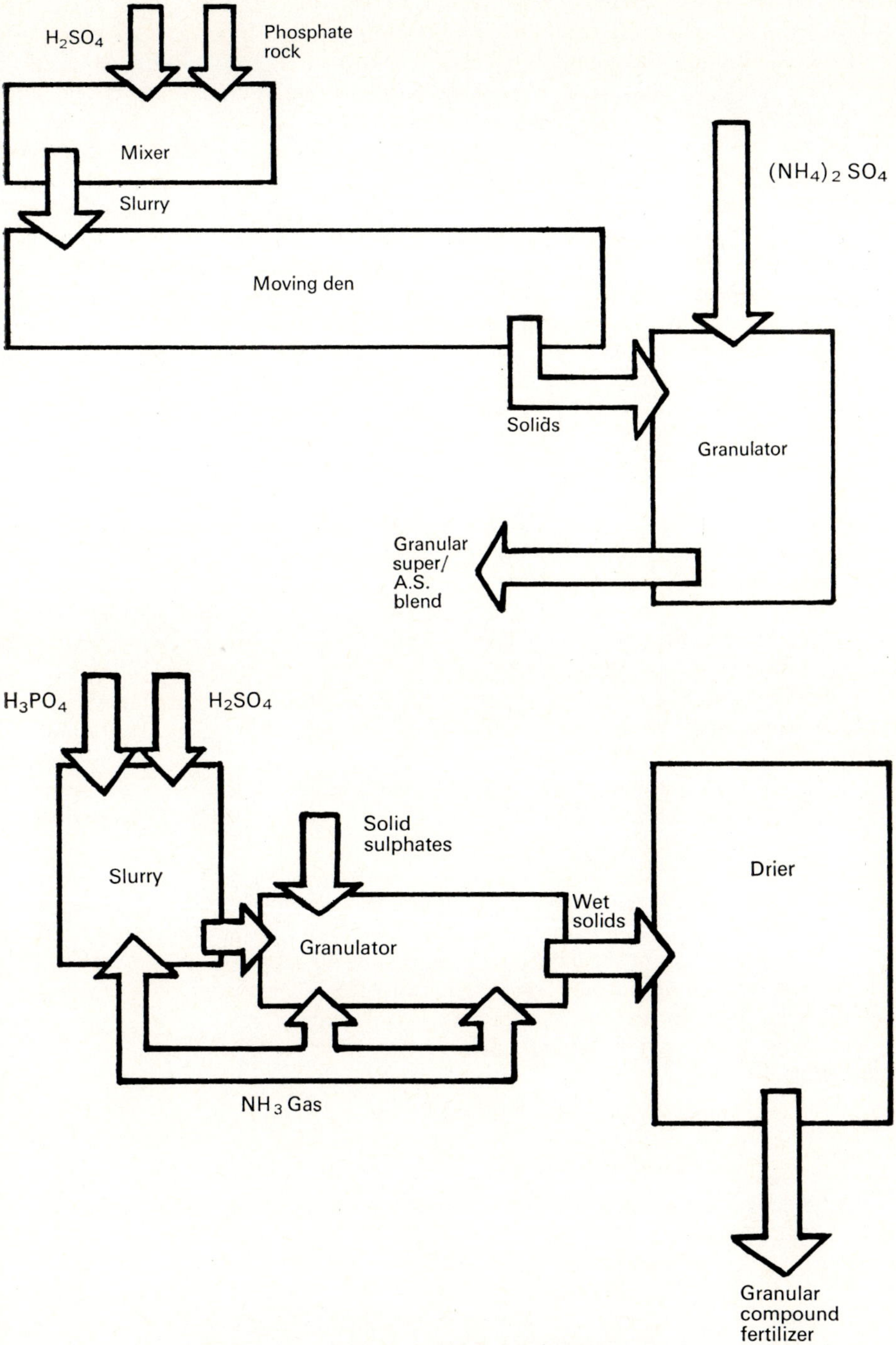

Figure 6-3 (a) Super/Sulphate of ammonia process (above).
(b) Ammonium Sulphate/Phosphate process (below).

either the sulphuric and/or phosphoric acids being ammoniated, or from solids such as potassium sulphate. The normal 'slurry' type granulation process for producing compound fertilizers is shown in Figure 6-3(b).

Turning now to examples of specially prepared high sulphur fertilizers, the 'sulphur-fortified superphosphates', the two grades marketed in New South Wales are known as 'S.F.25' and 'S.F.45'. They contain a mixture of sulphate and elemental sulphur, as shown in Table 6-1.

Initially, these products were made by simply preparing a physical mix of powdered superphosphate and sulphur which had been screened to about minus 6 mesh (3.5 mm). Naturally, the sulphur fraction contained a range of particle sizes from just minus 6 mesh down to very fine powder minus 100 mesh (0.15 mm). While the small particles are eminently suitable for reasonably quick nutrient release, the larger ones would take some years to become available. A considerable amount of experimental work was necessary to obtain Department of Civil Aviation approval for aerial spreading of these products, but eventually it was found that the fine superphosphate of that period effectively 'blanketed' the fine sulphur particles from explosion hazards.

Next, solid sulphur was added to the mixer when the phosphate rock was acidulated with sulphuric acid. The high temperature generated by the acid-rock reaction brought about a phase change in the sulphur, which adequately dispersed it through the fresh superphosphate, and gave a very homogeneous mix. This product was an improvement on the old 'dry mix' method, and the process is illustrated in Figure 6-4(a).

Following upon this development, equipment was installed to accurately meter molten sulphur to the acidulation mixer, and this again gave an improved product.

Since the introduction of a policy of granulation of all superphosphate, a new process for making these products has been developed and is illustrated in Figure 6-4(b).

Most of the manufacturing problems associated with S.F. supers now appear to have been overcome and there are now available two grades of Sulphur Fortified Superphosphates for treating areas deficient in both P and S. These mixtures are homogeneous, containing both sulphate sulphur for fast release and elemental sulphur for longer term release, and are now presented in a granular form for ease of handling and spreading.

At present three major products are used in the top-dressing areas and their P:S ratios are:

Single Superphosphate	1:1.3
S.F.25	1:3.5
S.F.45	1:7.6

In the cropping areas, the major products used in New South Wales and their P:S ratios are:

Double Superphosphate	3.5:1
Triple Superphosphate	20:1
Mono Ammonium Phosphate	15:1
15:13:0 (Starter 15)	1.3:1
11:15:9 (Grower 11)	3:1

Figure 6-4

(a) Sulphur fortified super, process A (above).
(b) Sulphur fortified super, process B (below).

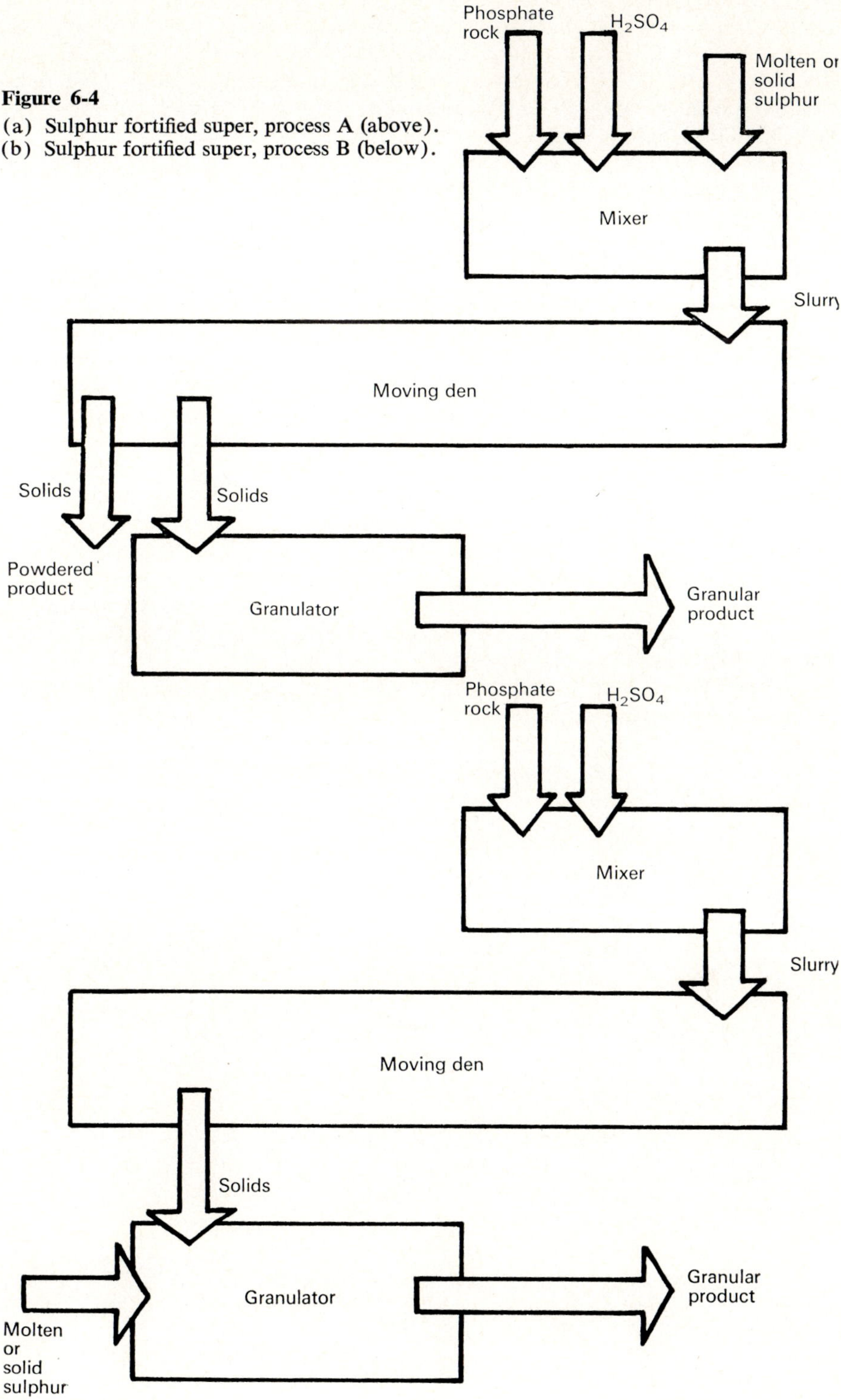

From this, it seems that the cropping situation is quite well catered for, particularly when it is considered that there are other NP and NPK products with a range of P:S ratios from 0.2:1 to 6:1. In addition, the more popular products such as double superphosphate and 'Starter 15' provide a very good insurance against sulphur deficiency.

While no doubt many combinations of products could produce various P:S ratios, there are a lot of hidden problems associated with their manufacture and handling and all these variations from the standard range of products would cost a good deal more to produce.

Already, overseas, completely new type sulphur-containing fertilizers are being produced. Sulphur is being coated onto urea and incorporated with diammonium phosphate. Elemental (preferably molten) sulphur is being dissolved in anhydrous ammonia, its solubility increasing with a decrease in temperature. It is applied to the soil using anhydrous ammonia equipment with only minor modifications. Usually about 10 per cent by weight is dissolved. In the soil, the sulphur is precipitated in a very finely divided form which oxidizes rapidly and thus becomes available very quickly. Ammonium polysulphides can be applied directly to the soil, metered into irrigation water, or mixed with anhydrous ammonia or ammonia solutions, which are in turn applied to the soil. They usually contain 20 per cent N and 40 to 45 per cent S. Ammonium thiosulphate is another material gaining in popularity. It can be applied in irrigation water and is compatible with many fertilizer solutions, such as aqua ammonia and various nitrogen solutions. It usually contains about 12 per cent N and 26 per cent S.

The call for such specialty fertilizers will be prompted by the development of more intensive farming practices, and by more precise knowledge of plant and soil requirements but, of course, the rate and extent of development of such materials will still be dictated by the prevailing economies.

REFERENCES

(1) *Looking Ahead With Sulphur*. Texas Gulf Sulphur Company, New York.
(2) Weir, R. G. (1967). *Agric. Gaz. N.S.W.*, **78**: 486.

Biosuper—Biological Superphosphate

R. J. SWABY

Biosuper, is a biological form of superphosphate, which relies on thiobacilli to oxidize elemental S to H_2SO_4 that in turn partially dissolves rock phosphate. This microbial fertilizer was only moderately successful on temperate pastures with low winter temperatures. However, higher soil temperature in the tropics during the wet season should favour thiobacilli, and it was decided to compare biosuper with super. Previous field tests with biosuper by Fisher and Norman (2) at Katherine and by Austin and Swaby (unpublished) at nine other sites in the Northern Territory had given promising results. Most soils in north Queensland are deficient in P and some in S so both elements are often needed for establishing and maintaining Townsville Stylo (Stylosanthes humilis). Rock phosphate-S fusions, as used by Attoe and Olson (1) and Nimgade (3), were not included because when compared with biosuper in pot trials they were inferior.

MATERIALS AND METHODS

Under our instruction, Australian Mineral Development Laboratories, produced biosuper pellets (av. 4 mm diam.), containing 16.6 per cent of S and 18.5 per cent or 26.1 per cent P_2O_5. Five parts by weight of ground rock phosphate (68 per cent $<$ 100 mesh per 2.5 cm) from Lady Annie deposits were dry mixed with one part of powdered elemental S (99 per cent $<$ 200 mesh), soil inoculum of thiobacilli (0.1 per cent) added, then all sprayed with tap water (approx. 8 per cent) on a pelletizing disc (1 m diam.), inclined at 44° and rotating at 19 rpm. At twelve sites in north Queensland biosuper was compared with superphosphate and often with rock phosphate, plus and minus gypsum and with untreated control. They were broadcast over small plots (0.004 ha, random blocks triplicated) by plane, and/or over long strips (.81 ha, random blocks duplicated) by plane or broadcaster. Equal amounts of P were added to the soil with each fertilizer. Biosuper was also applied at 1.5 and/or 2 times this amount, since the estimated cost of biosuper is much less than that of super. Equal amounts of S also were applied in superphosphate and rock phosphate plus gypsum, but the three rates of application of biosuper provided 1.5, 2.25 or 3 times this amount of S. At sites with low annual rainfall (460-635 mm) and with medium (635-1015 mm), superphosphate was applied at 251 kg/ha, but with high rainfall ($>$1015 mm) the rate was doubled. Fertilizers were applied only in the first year so that residual effects could be determined over a period of three years. Plot trials were never grazed, but mowings were removed each year. Cattle grazed the strip trials during the first-half of each growing season, then stock were ex-

TABLE 6-2
Effect of fertilizers on pasture grown in long strips at twelve Queensland sites

Site	Rainfall	Harvests	Mean Dry Weight, kg/ha Super-phosphate	2 Bio.	Bio.	R.P.+Gyp.	R.P.	Gyp.	Control	L.S.D. (P=0.05)
Antill Plains	Low	All	6,120*	6,870*	5,380	4,800	4,750	4,460†	4,630†	1,470
Charters Towers			4,460*	3,890*	3,720*	3,180†	3,040†	2,630†	2,420†	840
Fanning River			3,780*	3,330	2,890†	2,510†	2,700†	3,410*	2,540†	800
Georgetown			2,890*	4,500*†	3,000*	2,570*	2,830*	2,400	1,710†	750
Normanton			2,760*	2,480*	2,360*	1,640†	1,670†	1,370†	1,410†	620
Mount Garnet	Medium		1,070	1,480*†	1,380*†	1,070	1,000	950	830	290
Bowen		1st	2,620*	2,980*	2,930*	2,280			2,010†	440
		2nd	4,810	5,870	4,990	4,450			4,110	N.S.
Christmas Creek		1st	2,600	2,420	2,500	2,310			1,900	N.S.
		2nd	1,890	1,870	2,000	1,490			1,450	N.S.
Lakeland Downs		1st	3,710	3,720	4,020	3,300			2,870	N.S.
		2nd	1,830	2,360	2,400	1,750			1,220	N.S.
Cowley Beach	High	1st	1,010	1,060	1,250				1,100	N.S.
		2nd	7,870	8,920*	9,430*†				7,170	1,230
Malanda		1st	1,940	2,420*†	2,130*				1,730	290
		2nd	1,860	2,720	2,020				1,700	N.S.
Yuruga		1st	6,180*	6,360*	6,400*				3,580†	1,570
		2nd	6,920*	9,500*†	8,040*†				2,920†	890

* Values which are significantly different from Control.
† Values which are significantly different from Superphosphate.

cluded to permit regrowth. Green hay weights were recorded, and subsamples were taken for determination of moisture, P and S (X-ray fluorescence). Dry weights of herbage, uptakes of P and S and increments were calculated.

RESULTS AND CONCLUSIONS

Table 6-2 sets out the mean dry weights of herbage obtained from the strip trials. Space does not permit publishing the extensive data for uptakes of P and S from the strip trials, or any data at all from the plot trials.

Harvest counts

An attempt was made to assess the relative merits of each fertilizer applied on plots and strips by counting the number of harvests at each site where the dry weight of herbage, also the phosphorus and sulphur uptake values from each treatment, equalled or exceeded those from superphosphate. These counts were converted into percentages of harvests which met the criteria both from the crude data and also from the data after statistical analysis.

Without exception, the percentages based on the crude data were lower than the corresponding values after statistical analysis. While plots and strips mostly agreed about the ranking and magnitude of the percentages derived from the statistical data they disagreed with those from the rough data. It is considered that the data from the strips are more applicable to commercial pastoral conditions, therefore data from strips only are discussed.

The crude figures suggest that biosuper might be expected to equal or outyield superphosphate for dry weights in 72 per cent of cases, for P in 22 per cent and for S in 61 per cent. The corresponding percentages after statistical analyses were 94, 83 and 100. Usually double rates of biosuper gave higher percentages than single rates, while rock phosphate and gypsum gave lower percentages. The relatively high percentages, after biometry, from the control treatment, viz., 56, 56 and 72 suggest that this method of scoring the relative merits of fertilizers seems unsatisfactory. Analyses of increments were tried next, since control values are first subtracted from the results of each fertilizer.

Increment analyses

This second method of comparing fertilizers led to Table 6-3 which shows the mean incremental values of dry weight of herbage, and uptakes of P and S resulting from each fertilizer applied to strips only. These values have been expressed also as percentages relative to superphosphate increments. In order to show the effect of rainfall on these increments and percentages, the data were grouped into sites with low rainfall (5 sites), medium (4) and high (3). The treatment 1.5 biosuper was carried out at two of the three high rainfall areas, necessitating a second column of values for comparison.

The incremental weights were highest from the sites receiving most rain, and lowest from those with medium rainfall. However, the percentages relative to superphosphate deserve more attention and these were greater from the high and medium rainfall sites than from the drier areas. Thus biosuper gave incremental weights of herbage 146, 151 and 64 per cent of the superphosphate values at sites with high, medium and low rainfall respectively.

Double rates of biosuper gave even higher weight increments, which were 168, 167 and 114 per cent of the superphosphate figures, but rock phosphate plus gypsum gave much lower percentages. The corresponding percentages for P uptake increments were 115, 60 and 46, showing that biosuper provided more P than superphosphate under high rainfall conditions, where leaching might occur, but provided less under medium or low rainfalls. Once again double

TABLE 6-3
Effect of rainfall on pasture increments from long strips at twelve Queensland sites

Fertilizer increment	Increment (kg/ha) or % of (Sup.-Con.)	Low rain (5)	Medium rain (4)	High rain (3)
Dry weight data				
Sup.-Con.	Weight	1,460	510	1,270, 1,800
	percentage	100	100	100, 100
2 Bio.-Con.	Weight	1,670	850	2,130, 2,770
	percentage	114	167	168, 154
1½ Bio.-Con. [2]	Weight			2,710
	percentage			150
Bio.-Con.	Weight	930	770	1,850, 2,590
	percentage	64	151	146, 144
(R.P.+Gyp.)-Con. [9]	Weight	400	309	
	percentage	27	59	
Phosphorus data				
Sup.-Con.	Weight	1.23	0.89	4.09, 5.48
	percentage	100	100	100, 100
2 Bio.-Con.	Weight	0.97	0.80	4.86, 6.32
	percentage	79	90	119, 115
1½ Bio.-Con. [2]	Weight			6.87
	percentage			125
Bio.-Con.	Weight	0.57	0.53	4.72, 6.83
	percentage	46	60	115, 124
(R.P.+Gyp.)-Con. [9]	Weight	0.28	0.26	
	percentage	23	29	
Sulphur data				
Sup.-Con.	Weight	1.89	1.02	2.21, 3.14
	percentage	100	100	100, 100
2 Bio.-Con.	Weight	1.90	1.66	3.66, 4.51
	percentage	101	163	166, 144
1½ Bio.-Con. [2]	Weight			4.78
	percentage			152
Bio.-Con.	Weight	1.34	1.17	3.33, 4.50
	percentage	71	115	151, 144
(R.P.+Gyp.)-Con. [9]	Weight	0.63	0.56	
	percentage	33	55	

rates of biosuper gave higher results than single. Rock phosphate + gypsum gave the lowest percentages of P uptake relative to superphosphate. The corresponding percentages for S uptake increments were 151, 115 and 71, but it should be noted that biosuper contains 50 per cent more S than super. The availability of S from biosuper depends on two reactions, viz., rate of oxidation of S to sulphate and rate of leaching of sulphate from sandy soils. Both reactions are probably increased by rainfall. Biosuper, by supplying sulphate steadily, should have some advantages over superphosphate for sandy soils in high rainfall areas, but the results suggest that biosuper was no more effective than superphosphate as a source of sulphur, but was much more effective than rock phosphate plus gypsum. Double quantities of biosuper were better sources of S than superphosphate. As a low cost fertilizer, biosuper would be more economic for increasing dry matter of tropical pastures and as a source of P and of S at high and medium rainfalls, but not at low rainfalls. Commercial biosuper would contain 26.7 per cent P_2O_5 and 16.6 per cent S, while super contains 22 per cent and 10 per cent respectively, therefore, freight per unit of P and S for biosuper would be considerably less than for superphosphate.

Residual effects

Slow acting fertilizers, like biosuper, might be expected to have better residual values than soluble superphosphate so the increments for dry weight, and for P and for S uptakes obtained from the various fertilizers were set out for each year and for all years in Table 6-4. Judging by the dry weight data, all fertilizers performed best in the third year, moderately in the second and least in the first year, but seasonal differences could have caused this as much as fertilizers. Compared with superphosphate the herbage weights from biosuper were lower in the first and third years and higher in the second, indicating no greater residual effects from biosuper. Twice the rate of biosuper gave greater dry weight increments than superphosphate in the first and second years and equal values in the third, whereas rock phosphate plus gypsum was not even one third as effective as superphosphate. From the P uptake increments, superphosphate supplied more P than any other fertilizer in the first year, similar amounts at both rates of biosuper in the second year and higher quantities in the third year. Once again there was no consistent evidence for greater residual P from biosuper. Likewise the S data does not indicate consistently better residual effects from biosuper. Superphosphate and biosuper supplied the same total amounts despite the higher S content of biosuper.

Although the merits of different fertilizers can be compared by determining total uptakes of P and S by pastures, nevertheless the nutritive quality of the herbage from each fertilizer, particularly the contents of P and S, is of greater importance to grazing animals. At two sites grasses were hand separated from legumes and analysed separately. Without exception, the contents of P and S were higher in the legume components, but there were scarcely any significant differences between the contents from the various fertilizers. At the remaining ten sites the mixed swards were analysed for percentages of P and S and these data appear in Table 6-5.

TABLE 6-4
Residual effect of fertilizers on pasture increments from long strips at twelve Queensland sites

Fertilizer increment	Increments (kg/ha)			
	Year 1	Year 2	Year 3	Total
Dry weight data				
Sup.-Con.	960	1,110	1,540	3,610
2 Bio.-Con.	1,290	1,730	1,530	4,550
Bio.-Con.	820	1,220	1,250	3,290
(R.P.+ Gyp.)-Con.	280	360	500	1,140
Phosphorus da a				
Sup.-Con.	2.01	2.07	1.18	5.26
2 Bio.-Con.	1.80	2.37	0.98	5.15
Bio.-Con.	1.35	2.15	0.73	4.23
(R.P.+ Gyp.)-Con.	0.23	0.29	0.27	0.79
Sulphur data				
Sup.-Con.	1.54	1.73	1.74	5.01
2 Bio.-Con.	1.95	2.63	1.82	6.40
Bio.-Con.	1.61	1.96	1.59	5.16
(R.P.+ Gyp.)-Con.	0.71	0.60	0.55	1.86

Percentages of P and S

Non-breeding cattle need a minimum of 0.15 per cent P and 0.10 per cent S in their diet. Superphosphate almost met this criterion for P in the first year only and for S in the first and second years. Biosuper did not give an adequate percentage of P in any year, but met the dietary requirements for S in the first and second years. Double rates of biosuper gave higher contents of P than biosuper, but not quite as much as super in most cases. Rock phosphate plus gypsum was inadequate for P and acceptable for S only in

TABLE 6-5
Mean percentages of phosphorus and sulphur in pasture from long strips at ten Queensland sites

Treatment	% Phosphorus			% Sulphur		
	Year 1	Year 2	Year 3	Year 1	Year 2	Year 3
Super	0.148	0.094	0.062	0.153	0.110	0.091
2 Bio.	0.127	0.080	0.051	0.165	0.116	0.088
Bio.	0.109	0.078	0.048	0.152	0.108	0.092
R.P.+ Gyp.	0.060	0.046	0.046	0.110	0.092	0.079
R.P.		0.037	0.045		0.079	0.072
Gyp.		0.037	0.043		0.085	0.080
Control	0.101	0.067	0.042	0.141	0.096	0.069

the first year. When biosuper is used, mineral licks might be needed to augment the dietary intake for P, but not for S.

Since rainfall might well affect the percentages of P and S in herbage, Table 6-6 sets out the relevant data. Under high rainfall conditions the herbage from the various fertilizers, and even for the controls, were satisfactory both for P and S. Under conditions of medium rainfall, and particularly under low, the herbage was deficient for P even from superphosphate, but more so from biosuper. Superphosphate and both rates of biosuper provided adequate S contents in the herbage at high and medium rainfalls, but not at low rainfalls.

Faults and improvements

From observations made in the field during tropical downpours it appeared that pellets of biosuper broadcast on the surface were disintegrated by the bashing action of raindrops. When S particles become separated from rock phosphate much of the H_2SO_4 produced is dissipated reacting with soil constituents instead of with rock phosphate, thus decreasing the availability of P. Biosuper buried below the soil surface slaked less than pellets broadcast and it performed slightly better in pot experiments. By passing these biosuper pellets through hot air at 150°C for three seconds the outer particles of S were melted sufficiently to stabilize the pellets so that they withstood 3300 mm/day of simulated tropical rain. In pot trials heat-stabilized biosuper performed well. Heat-stabilized pellets survived transport over bumpy roads and withstood the pressures developed at the bottom of storage heaps. Since S dust does not arise from sintered pellets during bagging and handling operations they are not irritant to workers. The slow initial availability of P from biosuper can be improved by adding 10 per cent of superphosphate. The Department of Civil Aviation permits aerial top-dressing with fertilizers containing up to 22 per cent of elemental S without risk of explosion, so biosuper containing 16.6 per cent S meets this requirement. Biosupers containing different ratios of rock phosphate to S have been tested but the most economic mixture is 5:1. Inoculating with thiobacilli accelerates the oxidation of S and release of P, particularly on soils deficient in native sulphur-oxidizers.

TABLE 6-6
Effect of rainfall on percentage of phosphorus and sulphur in pasture from long strips at ten Queensland sites

Treatment	High rainfall		Medium rainfall		Low rainfall	
	P	S	P	S	P	S
Super	.194	.173	.083	.101	.054	.092
2 Bio.	.172	.182	.064	.113	.046	.091
Bio.	.163	.182	.059	.098	.041	.088
R.P.			.045	.066	.040	.079
R.P. + Gyp.			.060	.095	.040	.086
Gyp.			.049	.074	.038	.085
Control	.144	.171	.055	.084	.033	.071

The problem of preserving non-sporing thiobacilli from death by dehydration on biosuper pellets during storage has not been completely solved. Classical methods of protecting bacteria by coating them with colloids like proteins, gum arabic, carboxy-methyl cellulose and peat were unsuccessful, only clays having some beneficial effects. Numerous other colloids are being tested. Biosuper receives no government bounty, like superphosphate, because it is only a potentially soluble form of phosphate, but it might qualify for a proportional bounty. It is doubtful whether biosuper can be patented since its method of manufacture has been published in reports. It is not yet produced commercially in Australia, but these results might encourage firms to prepare trial lots in the near future. It can be manufactured with simple machinery at any port where rock phosphate and sulphur are unloaded. Developing countries, which lack factories, plan to do so. It has never been tested on annual crops but field trials are being undertaken at present.

Biosuper is a fertilizer which could be used economically on cattle pastures in the wetter tropics of Australia and other countries. It is superior to superphosphate in areas exceeding 635 mm (25 inches) of rain for increasing the yield of herbage containing adequate dietary levels of S, but rather inadequate levels of P, so that phosphatic mineral licks would be necessary. Improved forms of biosuper aiming at greater stability of pellets, higher residual effects of P and S, longer viability of thiobacilli, and quicker release of P have been tested in pot experiments, but not yet in the field.

Acknowledgements. The CSIRO gratefully acknowledges financial and technical assistance received from many sources during the development and testing of biosuper.

REFERENCES

(1) Attoe, O. J. and Olson, R. A. (1966). *Soil Sci.*, **101**, 317.
(2) Fisher, M. J. and Norman, M. T. J. (1970). *Aust. J. Exp. Agric. Anim. Husb.*, **10**: 592.
(3) Nimgade, N. M. (1968). *Trans. IX Congr. Int. Soil Sci.* (Adelaide), **2**: 765.

Sulphur in New Zealand Fertilizers

F. B. MULLER, G. McSWEENEY and J. ROGERS

Superphosphate has been, and still is, the principal phosphorus fertilizer used in New Zealand. About four million hectares are topdressed in the North Island and about two million in the South Island. Diammonium phosphate, triple superphosphate, and other imports contain less than 0.5 per cent of phosphorus applied. Fortunately the calcium sulphate in superphosphate has prevented sulphur deficiency on free-draining New Zealand soils. Now that the significance of sulphur is recognized (2), widespread deficiencies caused by the use of fertilizers lacking sulphur are less likely to arise. Nitrogen, phosphorus and potassium are registered under the New Zealand Fertilizers Act, 1960, but not sulphur.

In 1956-7 (4) the fertilizer industry used 100,000 tonnes of sulphur in making superphosphate and elemental sulphur was first recorded as an additive, 3 tonnes being used in the South Island. By 1965-6 190,000 tonnes of sulphur were used in superphosphate manufacture with an additional 13,000 tonnes of elemental sulphur used in the South Island in sulphur fortified mixtures. In 1972-3, 2.1 million tonnes of superphosphate in a total 2.5 million tonnes of fertilizer supplied about 230,000 tonnes (155,000 North Island, 75,000 South Island) of sulphur. Elemental sulphur use increased to 19,000 tonnes, 3,000 tonnes being applied principally to free-draining pumice soils in the central plateau of the North Island. In addition, 6,200 tonnes of sulphur is annually applied as imported ammonium sulphate with small amounts as potassium, copper and other sulphates, approximately 60 per cent being used in the South Island.

Some manufacturers add liquid sulphur to the mixer during the acidulation of the phosphate rock with sulphuric acid, and the recrystallized sulphur is very finely subdivided in the homogeneous mixture which, when granulated, is resistant to breakdown during handling. In the 10 and 20 per cent elemental sulphur-superphosphate mixtures commonly manufactured, one-half and one-third respectively of the sulphur is sulphate. The S:P ratio ranges from about unity in superphosphate to about 3.5 in the 20 per cent sulphur mixture.

Sulphur in dry mixes is screened to pass 2.8.4.0 mm, about 50 per cent being minus 0.5 mm and 20 per cent minus 0.15 mm. Additions greater than 225 kg/t are not recommended as dry mixes because of explosion hazards, which may be reduced by granulation. Trial additions of larger amounts of molten sulphur into the superphosphate were not successful, the mixture remaining fluid too long in the den.

Australian work (7) showed that, when molten and solid sulphur were added to the mixer, over 60 per cent and 40 per cent respectively of it was finer than 0.15 mm in the superphosphate mixture whereas dry mixes contained only about 10 per cent of this fine fraction.

Winter rains often leach sulphate (1) particularly from soils with low-retentive capacities for sulphate when superphosphate is applied in the autumn. Granulation retards solution of calcium sulphate from superphosphate and from sulphur fortified superphosphate in which the elemental sulphur also is less mobile because much less surface is exposed for oxidation by bacteria.

Near Taupo at Lake Rotokaua (6) some six million tonnes of elemental sulphur is the largest deposit yet found in New Zealand. This and the pyrites ore bodies in the Thames and Nelson areas (3) are not yet developed commercially.

EXPERIMENTAL

A pot trial investigated the uptake of phosphorus (P) and sulphur (S) from granular and powder forms of superphosphate by New Zealand white clover on Oruanui sand and Hurunui loam, and the nutrient losses in the drainage. The Oruanui sand, a yellow-brown pumice soil (5) was probably not fertilized and was obtained from a mature pine plantation where annual rainfall averages 1,214 mm usually well distributed. The Hurunui stony silt loam—a yellow-brown earth obtained from Wairau Valley, Marlborough, where annual rainfall is about 900 mm—had possibly received a little superphosphate.

Earthenware crocks with centre drainage hole and holding about 4.5 litres of soil, 18.5 cm diameter, 17.5 cm deep with 269 cm^2 soil surface area were used for Hurunui soil; the corresponding measurements for Oruanui soil pots being 9 litres, 22.4 cm, 23 cm and 394 cm^2.

Pots were free-draining and the same amount of leachate was collected from each pot which amounted to 25 mm from the Hurunui soil and 36 mm from the Oruanui for each five-weekly leach and harvest cycle.

Each treatment was replicated four times. The basal treatment (5) (Table 6-7) was 800 kg/ha superphosphate and 200 kg/ha KCl. Treatments 5 and 6 were with the same superphosphate sampled at den cut-out, and after granulation; while Treatments 7 and 8 were made by FMRA with the superphosphate used in Treatment 5. Treatments 9 and 10 were made at the Bay of Plenty Co-operative Fertilizer Company, Treatment 9 being ground to pass a 1 mm sieve.

The P and S in the treatments were balanced by adding elemental S (powder or granules) and monocalcium phosphate (MCP).

The clover was sown on 29 March 1972, and after taking off one crop in late May (Harvest 0) the Hurunui soil was topdressed on 1 June, and the Oruanui on 6 June. Harvest 10 was obtained from the Hurunui soil on 27 June and from the Oruanui on 2 July 1973. Major and minor nutrients (K, Mg, Ca, Cu, B and Zn) were applied (Figures 6-5 and 6-6) as herbage analysis intimated actual or likely deficiencies.

RESULTS AND DISCUSSION

Although there was a quicker response in the first two harvests (Figures 6-5 and 6-6) with the powders, granulation usually increased total yields

TABLE 6-7

Dry matter obtained in 10 harvests and sulphur removed by cropping and by leaching

	Dry matter g/pot		S mg/pot		S in 11 leachates mg/pot		Percentage of applied sulphur removed					
							Cropping		Leaching		Total	
	Hurunui	Oruanui	Hurunui	Oruanui	Hurunui	Oruanui	Hurunui	Oruanui	Hurunui	Oruanui	Hurunui	Oruanui
1. Control	39.7	13.1	62	30	4	18	—	—	—	—	—	—
2. Gypsum	58.2	16.0	148	44	81	108	40	4	36	29	76	33
3. Elemental sulphur	57.6	12.7	131	31	6	23	32	0	1	2	33	2
4. Monocalcium phosphate	37.9	55.2	62	129	4	11	—	—	—	—	—	—
5. Superphosphate powder	68.2	64.2	163	159	123	44	43	8	50	10	93	18
6. Superphosphate granules	72.1	62.6	182	167	42	11	52	11	17	0	69	11
7. Dry S-fort super powder	72.5	62.5	229	171	272	115	20	5	33	13	53	18
8. Dry S-fort super granules	74.0	65.3	224	173	194	11	21	6	24	0	45	6
9. Molten S-fort super powder	72.3	59.3	235	170	457	207	21	5	57	23	78	28
10. Molten S-fort super granules	74.7	65.1	210	178	105	14	18	5	13	0	31	5
11. Serpentine superphosphate	66.9	61.7	170	152	108	44	43	6	41	9	84	15
12. Lime superphosphate	68.2	68.4	163	169	100	27	47	13	44	5	91	18
LSD at 5%	10.3	9.8										

Notes: Treatments 2, 3, 4, 5, 7, 9, 11, 12 (< 1 mm).
Treatments 6, 8, 10 ($> 1 < 4.75$ mm).

(Table 6-7). These results generally agree with Williams (8) who found that particle size influenced yield of dry matter, P and S uptake, and loss by leaching. At rates of 125 and 250 kg/ha superphosphate Williams found in one instance coarser granules decreased yield, whereas we find with 800 kg/ha superphosphate (a rate used in New Zealand for soil development) yields were not reduced by granulation.

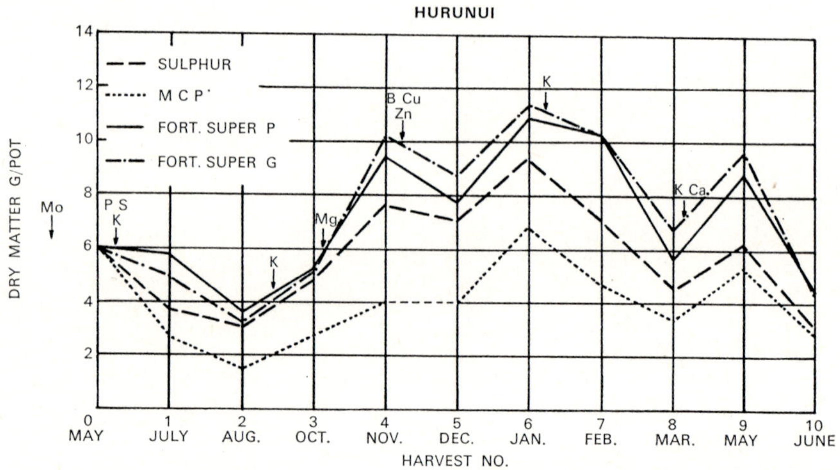

Figure 6-5 Yields of white clover during 1972-3 season from Hurunui stony silt loam treated with elemental sulphur, MCP = monocalcium phosphate, Fort super P = molten sulphur fortified superphosphate powder, Fort super G = molten sulphur fortified superphosphate granules.

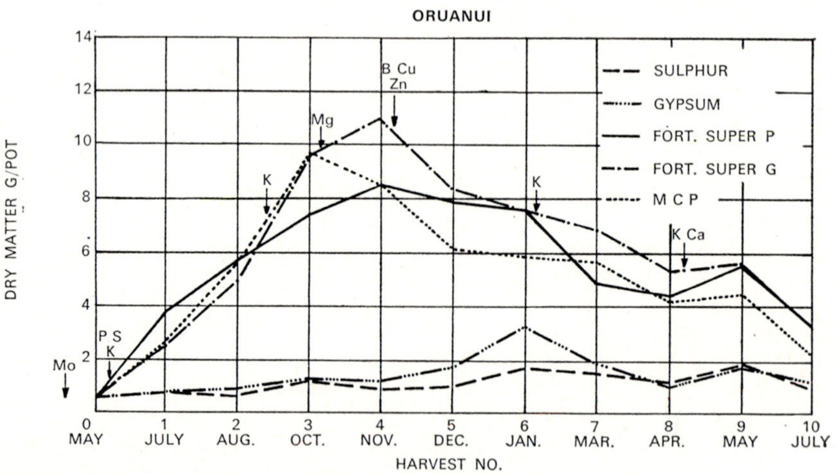

Figure 6-6 Yields of white clover during 1972-3 season from Oruanui sand treated with elemental sulphur, gypsum, Fort super P = molten sulphur fortified superphosphate powder, Fort super G = molten sulphur fortified superphosphate granules, MCP = monocalcium phosphate.

Hurunui

Clover yields obtained with the application of P alone did not differ significantly from control (Table 6-7 and Figure 6-5). S was not released rapidly enough from this soil to give optimum yields when other nutrients were added although it was well supplied with total S as well as P. Yields obtained with S alone, either as gypsum or elemental S were significantly better. The yield from gypsum was significantly higher than elemental S in Harvest 1 after which there was no difference.

The best yields were obtained with fertilizers containing both P and S, both powdered and granulated, with lower yields after Harvest 3 from the materials lacking P and significantly better than the S and P only materials. By repeating the experiment in 1973 the decline in yields in Harvest 2 and 3 is attributed to lack of K and possibly Mg.

Oruanui

The low yield of the control treatment, despite the 1.45 times greater area for clover growth in the larger pots, reflects the low fertility of this soil. Whereas Hurunui soil was S responsive, Oruanui was grossly P deficient. Yields from the two S treatments were not significantly different from control or from one another. In Harvests 5 and 6 gypsum gave greater yields, probably due to the lack of Ca in this soil and possible acidification from the S. Limestone was applied in April 1973.

There was no difference between MCP and the S-fortified superphosphates in Harvests 1 and 2 (Figure 6-6). From Harvest 3 on, the granulated material often was significantly better although the total yields were not. MCP yields fell in relation to the P + S fertilizers showing that soil S became depleted fairly early.

A reason for the lower yields from the powdered than from the granulated molten-S addition superphosphate on this Ca-deficient soil, may be the large surface of finer elemental S exposed to biological oxidation resulting in the formation of sulphuric acid which adversely affects the *Rhizobia*.

Use of P and S by clovers

Figure 6-7 and Table 6-7 show the high levels of P in, and the low yields of, the clovers grown on S-responsive Hurunui soil when MCP was used rather than superphosphate, relative to levels and yields with S only. With P deficient but S adequate, as on the Oruanui soil, the small amounts of clover grown without P had much lower levels of P which increased with both MCP and superphosphate. The clover used 24 per cent of the P in MCP and 36 to 40 per cent in the fertilizers containing both elements on Hurunui soil, and on Oruanui soil about 42 per cent of the P in all the phosphates.

Figure 6-7 also shows less than 0.2 per cent S in clovers grown on Hurunui soil without S fertilizer whereas very high levels were obtained with S-fortified superphosphate. On this soil, in which apatite is a source of P, S levels from gypsum, elemental S and superphosphate were similar early in the trial. Clovers grown on Oruanui soil did not have such low levels of S and fertilizer S did not increase plant levels so markedly.

The Hurunui soil provided only 25 to 30 per cent of the S used by the

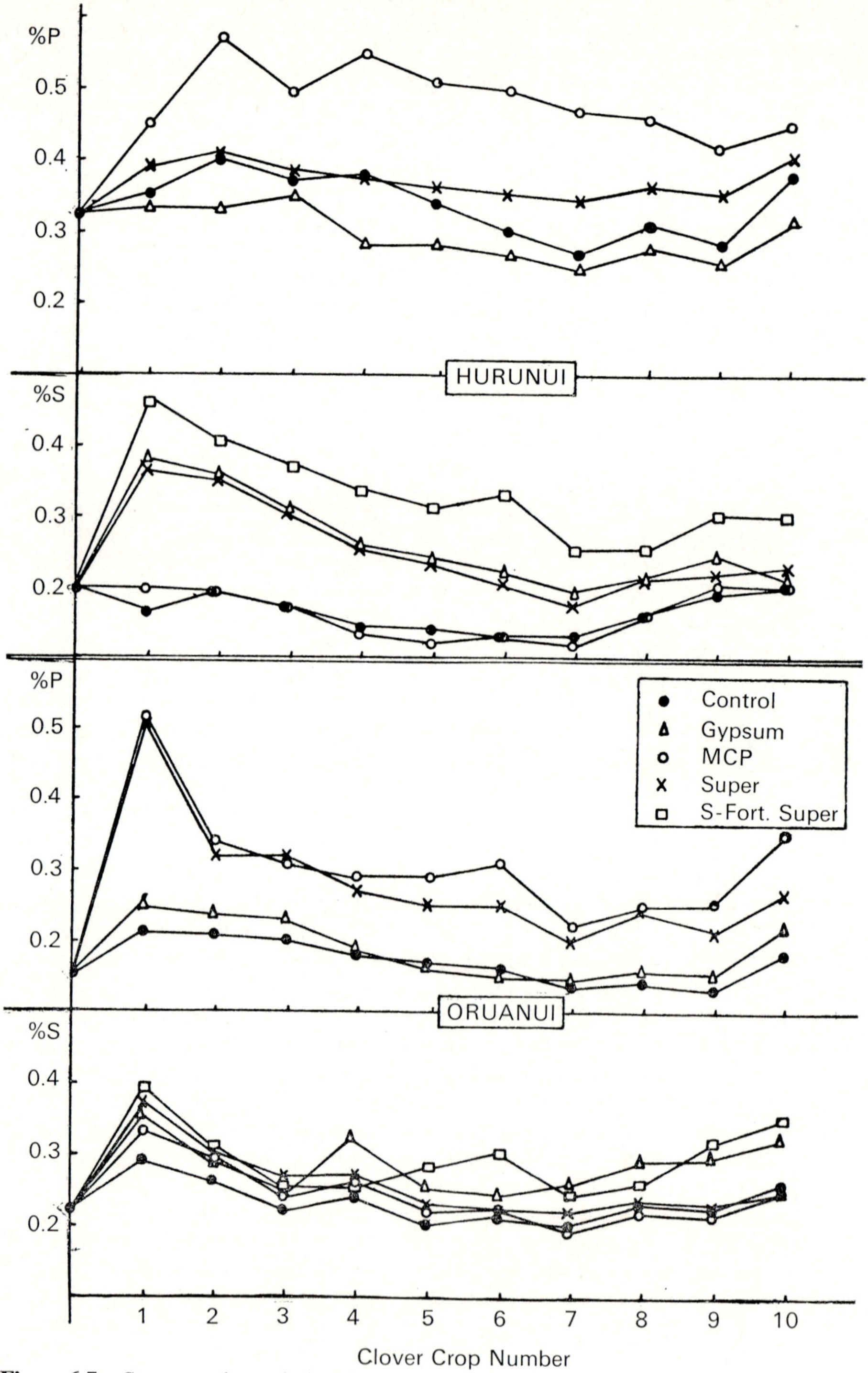

Figure 6-7 Concentrations of phosphorus and sulphur in successive clover crops from Hurunui and Oruanui soils.

clovers whereas Oruanui soil provided 70 to 80 per cent. The rest came from the fertilizers. On Hurunui soil 45 per cent and 20 per cent of the S in superphosphate and S-fortified superphosphate respectively was used by the plants and on Oruanui soil 10 per cent and 5 per cent.

Leaching losses of P and S

No leaching of P reflected the high P-retention capacity of these soils. Although calcium sulphate is less soluble than MCP, sulphate was readily leached particularly from Hurunui soil. The gypsum, superphosphate and S-fortified superphosphate powders lost considerable S on Oruanui soil whereas elemental S powder and the granular fertilizers did not. The loss of S from control was higher from Oruanui soil than from Hurunui soil confirming the soil analysis that Hurunui had little available S (Tables 6-7 and 6-8).

Figure 6-8 shows leaching differences were greater on the Hurunui soil. The highest concentrations were in the leach water from the S-fortified superphosphate powders, particularly the one made by adding molten S during acidulation wherein the S is inferred to be much finer than in Treatment 3 (< 2.8 mm) or in Treatment 7 (< 1 mm), so that it is more readily oxidized. On Hurunui soil superphosphate lost more S than gypsum possibly because of the preferential adsorption of phosphate ions and/or because the calcium sulphate is finer in superphosphate. On Oruanui soil without P there was insufficient growth to use available sulphur and more S was leached from gypsum than superphosphate.

Granulation reduced leaching losses of sulphur from both soils without adversely affecting plant yields. On Hurunui soil leaching losses from granulated superphosphate fortified with molten S were less than from the granular dry-mix product, indicating they were bonded by the S into a less soluble matrix.

Powdered superphosphate lost 50 per cent (17 per cent) of its S by leaching from Hurunui soil, which combined with 43 per cent (52 per cent) used by the clover left 7 per cent (31 per cent) for future plant growth. The brackets refer to granulated superphosphate (Table 6-7). Granulation of the molten addition S-fortified mixture reduced leaching losses from 56 per cent to 13 per cent.

Reversion of the MCP in superphosphate with serpentine, or lime, to less soluble magnesium and dicalcium phosphates on Oruanui soil reduced the S leached from 10 per cent to 5 per cent with lime and on Hurunui soil from 50 per cent to 41 percent with serpentine and to 44 per cent with lime.

To conclude, on a sulphur-responsive Hurunui soil application of phosphorus alone, and on the phosphorus deficient Oruanui soil sulphur alone did not increase plant growth. Use of sulphur alone to correct sulphur deficiency on Hurunui soil reduced yields in time due to insufficient phosphorus. the converse applied on phosphorus deficient Oruanui soil.

On these soils with low sulphur retention capacity there were losses of sulphur by leaching from powdered fertilizers. Granulation reduced sulphur leaching from superphosphate. Leaching from superphosphates containing elemental sulphur was also reduced by granulation or by using coarser sulphur.

TABLE 6-8
Chemistry of the soils and fertilizers used

	Soils		Fertilizers			
	Hurunui	Oruanui	Treatment		% P	% S
pH	5.4	5.8	1	Control	—	—
Sulphur* (mg % S): Adsorbed	0.3	5.0	2	Gypsum	—	18.6
total	55	17	3	Elemental sulphur		100
Phosphorus* (mg % P): Truog extractable	2	0.6	4	Monocalcium phosphate	24.4	—
N H_2SO_4 extractable	26	10	5	Superphosphate powder	9.9	10.6
Inorganic	35	21	6	Superphosphate granules	9.9	10.6
Organic	91	4	7	Dry S-fort super powder	8.3	30.1
Total	126	25	8	Dry S-fort super granules	8.3	30.1
Cation exchange capacity (me %)	23.8	3.9	9	Molten S-fort super powder	8.4	29.1
Total exchangeable bases (me %)	15.8	1.3	10	Molten S-fort super granules	8.4	29.1
Per cent base saturation	66	33	11	Serpentine superphosphate	7.4	8.2
Exchangeable cations (me %): Ca	12.8	0.54	12	Lime superphosphate	7.4	8.5
Mg	2.2	0.10				
Na	0.29	0.25				
K	0.56	0.55				
Total	15.85	1.44				
Clay content w/w	3(illite type)	1(allophane)				
Per cent phosphorus retention	30	38				
Per cent sulphur retention	6	10				

* Sulphur and phorphorus retention by the MAF, and other soil sulphur and phorphorus measurements by the Soil Bureau, DSIR.

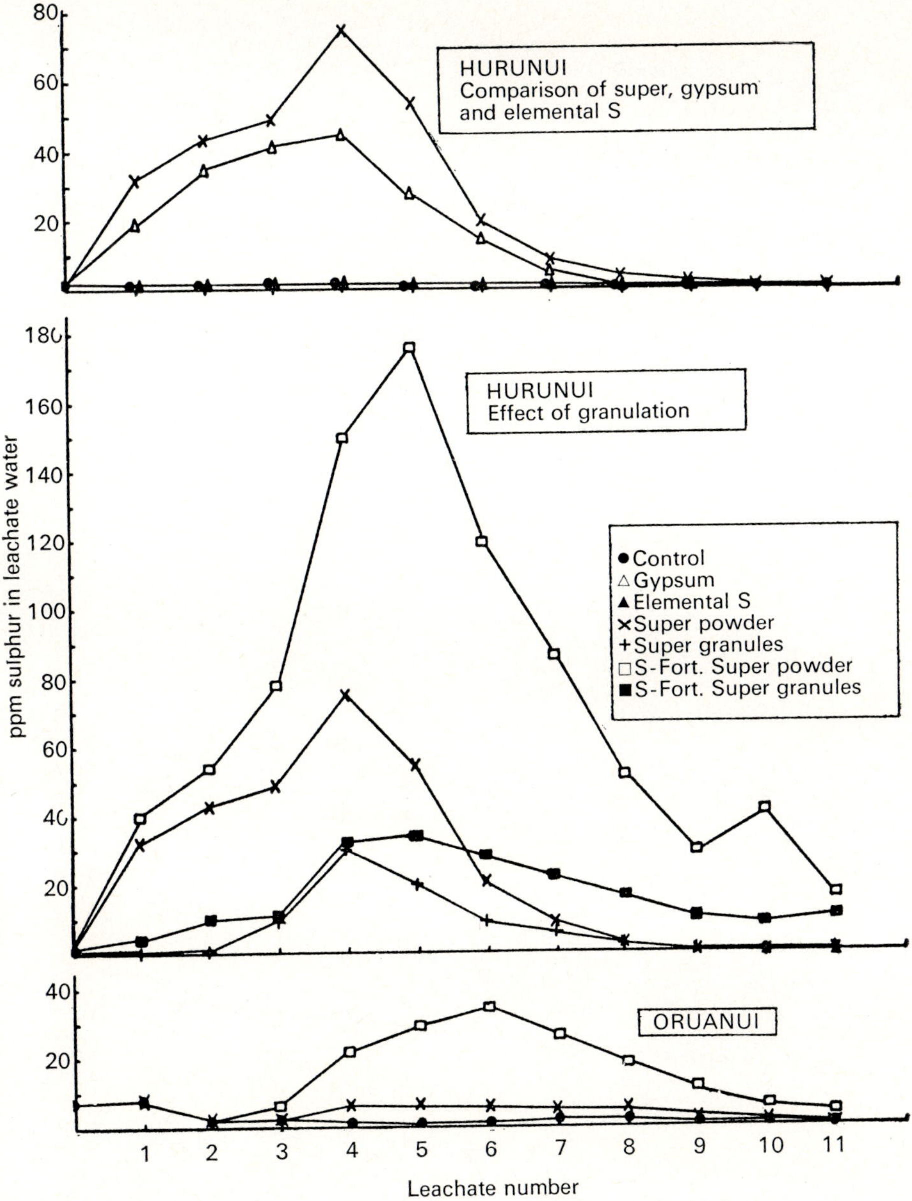

Figure 6-8 Sulphur concentrations in successive leachates from Hurunui and Oruanui soils.

REFERENCES

(1) Hogg, D. E. and Toxopeus, M. R. J. (1966). *N.Z. J. Agric. Res.*, **9**: 93.
(2) Ludecke, T. E. (1969). *Proc. XII Tech. Conf.*, NZFMRA: 85.
(3) Metson, A. J., Blakemore, L. C. and Chittenden, E. T. (1971). *N.Z. J. Sci.*, **14**: 104.
(4) N.Z. Ministry of Agriculture and Fisheries, Statistical Review of Fertilisers in New Zealand 1970/71: 19.
(5) N.Z. Soil Bureau (1968). 'Soils of New Zealand'. Bull. 26 Pt 3. Govt Printer, Wellington.
(6) O'Malley, G. B. (1969). *Proc. XII Tech. Conf.*, NZFMRA: 67.
(7) Weir, R. G. (1970). *Proc. XI Int. Grassld Congr.*, p. 380.
(8) Williams, C. H. (1970). *Proc. XI Int. Grassld Congr.*, p. 384.

Sulphur Resources Available to the Fertilizer Industry, the Fertilizers Made, Their Role, and the Problems Involved in Their Manufacture

D. W. BIXBY and V. J. KILMER

The earth's crust contains about 0.06 per cent S. Organic S in the soil becomes available as a nutrient with time, but supplementation by atmospheric or fertilizer S is necessary. While the fertilizer source can be controlled to meet agricultural needs, industry stack gases now provide a significant input to atmospheric S, and in the near future, much of it will be recovered and become a source material for fertilizer manufacture.

Plant nutrient S for fertilizers may be extracted in the elemental form from several natural sources. It may be obtained by Frasch mining methods from salt-dome deposits where it is present in the pure form. This use of geothermal water is being considered for extracting S from the Lake Rotokaua deposit in New Zealand. Elemental S may also be extracted by progressively more complex methods from sour gas, crude oils, and a number of coals.

Nature has also supplied abundant S in the form of metallic sulphides. Iron sulphide or pyrite is sometimes used to supply S specifically, but more often it is economically viable only if all values are recovered, e.g., the non-ferrous sulphides. These compounds are, at present, of little direct value as plant nutrients, but they may be employed to manufacture sulphuric acid.

Sulphates, the form of S utilized to the greatest extent by plants, also exist in nature. Gypsum, potassium sulphate, and potassium-magnesium sulphate are all present as evaporites from ancient sea beds. Natural gypsum is of little importance to the fertilizer industry as an S source because of its low nutrient content. Its main use is as a soil amendment. Potassium sulphate and potassium-magnesium sulphate have a long history as fertilizer sources of S, some incidental in the past, but increasingly deliberate in the present.

Complementing these natural sources is a long list of synthetic materials, some using elemental S, some sulphuric acid, and some sulphur dioxide or hydrogen sulphide to form compounds capable of supplying fertilizer sulphur. Chief among these are normal superphosphate, with its 'incidental' S, and ammonium sulphate. The former is declining in importance abroad, but the latter is gaining popularity due largely to an increasing awareness that it is a 45 per cent plant nutrient material, rather than 21 per cent. The same

TABLE 6-9
Sulphur-containing fertilizer materials

Sulphur as S	Sulphur as SO_4	Sulphur in other forms
1. Agricultural sulphur	1. Ammonium sulphate	1. Ammonium thiosulphate
2. Flake sulphur	2. Potassium sulphate	2. Ammonium bisulphite
3. Granular sulphur-bentonite	3. Calcium sulphate	3. Ammonium polysulphide
4. Phosphate rock-sulphur	4. Magnesium sulphate	4. Sulphur dioxide
5. Normal super + sulphur	5. Potassium-magnesium sulphate	5. Certain ultra high analysis materials
6. Triple super + sulphur	6. Normal and enriched superphosphate	6. Lime-sulphur
7. DAP + MAP + sulphur	7. Ammonium nitrate-sulphate	
8. Ammonium polyphosphate + sulphur	8. Ammonium sulphate-nitrate	
9. AP–UP + sulphur	9. Ammonium phosphate-sulphate	
10. NPK grades + sulphur	10. Ammonium phosphate-sulphate gypsum	
11. Urea-sulphur	11. Ammonium phosphate-sulphate urea	
12. Potash-urea-sulphur	12. Several micronutrient salts	
13. Sulphur coated potash	13. Urea sulphate	
14. Sulphur coated urea	14. Urea-ammonium sulphate	
15. Ammonia-sulphur	15. Sulphuric acid	
16. Sulphur suspensions	16. NPKS	
17. Micronutrients in sulphur		

reasoning applied to normal super makes it about 33 per cent plant nutrients, still too low to compete effectively in most of today's markets. Fortified with 180 kg S per ton, however, the number becomes a respectable 44 per cent.

There are more than 40 sulphur-containing fertilizer intermediates. These constitute the bulk of the S resources available to the fertilizer industry. Table 6-9 lists the principal S-containing fertilizer products. Most of these are well known and widely used. Prilled urea-ammonium sulphate is under development and may have a significant potential.

The problems, or lack of problems, of manufacture are, of course, inescapably bound up with the roles these products will play, although the reverse is not necessarily true. Consequently, these topics will be covered simultaneously.

ELEMENTAL SULPHUR

Elemental S by itself, as the agricultural, and to a lesser extent by the flake variety, is inherently dusty, and potentially hazardous. These problems may be overcome by particle size or prilling, but agricultural efficiency is reduced. Experimental quantities of porous sulphur granules have been made, but agronomic evaluations have not been reported.

The additive, bentonite clay, is used to manufacture a granular material, which disintegrates into finely divided S on wetting. Such a product is used both for mixes and direct application and should be suitable for aerial application. It is believed to be made by vat type cooling of a 90 per cent liquid S-10 per cent bentonite mixture, followed by a crushing operation. Other types of binders, such as ligno-sulphonates, have shown promise but are not yet commercially produced. The addition of soluble sulphate salts to these products provides an S source that is immediately available to plants.

SULPHUR SUSPENSIONS

Suspensions using elemental S are readily formulated. At TVA suspensions containing about 60 per cent S and 1 per cent clay were prepared from —20 mesh crushed lump sulphur. Using flowers of sulphur, suspensions of 40 per cent S were made. The greater viscosity of the latter reduces the amount of S which can be handled. Clay was first dispersed in water for 5 min., the S was added, and mixing continued for another 5 min. Complete formulas, such as 9-18-18-10 and 12-12-12-20, were also prepared using 12-40-0, urea-ammonium nitrate (UAN) solutions, potassium chloride (KCl), and sulphur.

A portable mixer suitable for field use using an engine-driven impeller blade (35 cm diameter) has been developed. Using crude S, a 60 per cent slurry is prepared, pumped to field applicators, and sprayed onto the soil using existing liquid fertilizer equipment. The S slurry can also be used as a vehicle for micronutrients.

Other means of suspension preparation are possible. Molten S can be dispersed in aqueous media, using techniques originally developed for pipelining sulphur. Best results were obtained with both S and solution under pressures of 4-60 psi. A more exotic method of making a stable S suspension involves reaction of H_2S and SO_2 in water containing a small amount of ethylene glycol.

SULPHUR COATINGS

Sulphur-coated materials which supply plant nutrient S deliberately have been introduced. These employ finely divided S bound with oil or formaldehyde to a granular product such as urea. A 40-0-0-10 grade was being marketed for use on rice and Coastal bermudagrass.

Several S-coated materials which supply plant nutrient S incidentally to their slow-release characteristics have been put forward, e.g., TVA's S-coated urea (39-0-0-10). ICI has recently developed a similar product in the U.K. Some S-coated potash has also been made at TVA. TVA has shown that urea coated with a smaller amount of S (3 to 4 per cent) and 0.5 per cent oil-wax is compatible with triple superphosphate. This is not normally the case. Although these products are really outside the scope of this paper, the expertise developed in the handling and spraying of liquid S for coating fertilizer materials should be of value in the making of materials where homogeneity is required. As one example, TVA is trying to reduce entrainment-caused S losses by experimenting with a hydraulic spray system which would replace the present air atomization setup.

PHOSPHATE-SULPHUR MATERIALS

A 20 psi hydraulic spray is quite satisfactory for adding S to most fertilizers in a drum or pan granulation system and was used for formulating a granule triple superphosphate S formulation (0-38-0-20). The product was enthusiastically received, but production problems, including explosions at the oversize grinding mill, caused them to shut down. The problems were believed curable with improved design and alternatives are being sought by the producer.

Sulphur fortification of normal superphosphate is widely and successfully practised in Australia and New Zealand. Originally, S screened to about 40 per cent —60 mesh was added to dry dressed superphosphate which was then passed through a ribbon mixer before dispatch. With the advent of granulation, additive segregation became a problem and a variety of other methods of adding S have been examined; viz., liquid and solid S added at the acidulation mixer, blunger, and granulation drum. The mixer has emerged as the preferred point of addition, with liquid S the preferred form for continuous production, and the fine solid for occasional operation.

Many production problems have existed, but most have been corrected. With solids, problems involved grinding, screening, removal of tramp iron venting, and segregation in the product. With liquids, filtering, pumping, metering, and fluctuation at cut-out of denned material have been troublesome. The latter problem may still exist, but the results of metering in reasonably clean S to the mixer using calibrated gear pumps appear to be quite satisfactory. Further, the S is so well incorporated that percentages much higher than the 500 lb/ton limit proposed in New Zealand for aerial (dry S-super mix) application can be tolerated. In Australia a fortified superphosphate product containing 45 per cent S added as a liquid, is routinely applied from the air.

Recent evidence (1) suggests that addition of liquid S at the mixer of a triple superphosphate manufacturing operation might be a more suitable means of incorporating S into that product for use in areas where a higher P_2O_5:S ratio is indicated.

Another phosphate-S material is phosphate rock-sulphur and more particularly 'Biosuper', a mix to which *Thiobacilli* has been added. Interest in such mixes goes back nearly sixty years but their agronomic value has still not been fully evaluated.

N-S MATERIALS

Molten S and urea are miscible, and a homogeneous granular product with excellent storage and handling properties can be made by prilling. A product (40-0-0-10) has been marketed and was reported useful in direct application, bulk blends, and granular processes.

TVA has produced granular urea S (about 44-0-0-4) by spraying molten S and urea melt as separate streams to the pan granulator in the pilot plant. The S was atomized to attain small particle size in the granular product.

The solubility of S in anhydrous ammonia has been exploited in the manufacture of a 74-0-0-10 grade. Production is straightforward, but corrosion and precipitation can be problems in application. Application equipment must be cleaned after use to avoid corrosion of valves and precipitation of S due to leakage of NH_3 from the pressurized system. With proper application techniques, including use of moisture-free ammonia, a vapour purge for the system, well sealed valve stem threads, and diligent mineral oil lubrication of moving parts in contact with the solution, this product is used on potatoes, onions and sugar beets.

N-P-S, N-K-S AND N-P-K-S MATERIALS

Elemental S has been incorporated in diammonium phosphate (DAP) and monoammonium phosphate (MAP) by metering it as one of several solid feed streams to a TVA ammoniator granulator, yielding homogenous products containing 12 to 15 per cent S. Other works have included up to 20 per cent S in MAP-DAP using pan granulation and blunger processes. Early season response is enhanced by including 15 to 20 per cent of the S in sulphate form.

By spraying liquid S on a rolling bed of granular ammonium polyphosphate, a grade of 12-52-0-15 was prepared by TVA, and is being evaluated in bulk blending operations.

A fertilizer based on ammonium phosphate, urea phosphate and elemental S (15-45-0-5) is described as particularly suited for use on alkaline soils in U.S. Patent 3,333,939. In U.S. Patent, 3,501,232, a fertilizer 35 per cent by weight of KCl and 5 to 40 per cent of a 1:1 mixture of urea and elemental S is described. It was aimed at upgrading finely divided potash and is said to have low hygroscopicity and low water solubility.

Two complete fertilizers analyzing 10-20-30-3 and 12-24-24-3 have been produced primarily for use on cotton. Elemental S is incorporated because of the crowded nature of the formula. Molten S was introduced to a TVA ammoniator granulator through spray nozzles above the bed.

In addition to the complete suspensions already mentioned above, suspension grades such as 4-12-24-1 and 5-18-18-1 are now routinely used, both for direct application and as base materials for use with 10-34-0 and ^{32}N solutions.

Micronutrient S fusions have been made by adding micronutrient salts to molten sulphur. Data suggest that prolonged release of Mn, Zn and Cu is possible with this system.

Production problems associated with solid products containing elemental S generally relate to the generation of potentially hazardous amounts of S dust. The hazard is present but may be less apparent when products dilute with respect to S are handled. However, appropriate equipment design, ventilation and good housekeeping can reduce or eliminate the potential danger.

SULPHATE-CONTAINING FERTILIZERS

Phosphorus—sulphate materials

An example is normal superphosphate. While its role is diminishing elsewhere, it remains important in Australasia due to its 10 to 14 per cent sulphur content, all in the available sulphate form. The incidental application of S in normal super has delayed the occurrence of S deficiencies in many areas of the world.

Nitrogen—sulphate materials

Ammonium sulphate remains popular. It now accounts for some 3.6 million tons of plant nutrient S, worldwide, and is expected to increase to 5.7 million tons in 1975-6, then taper off again. Of this, 58 per cent is currently synthetic, 16 per cent by-product and 26 per cent as a co-product from synthetic fibre manufacture. Its popularity has been due to its ability to compete in the N price area, through writing off certain costs against associated primary products. The S content is assuming increasing importance to consumers and is being charged for.

Ammonium nitrate-sulphate and sulphate-nitrate are names applied to grades containing AN and AS in varying amounts. A grade analysing 30-0-0-13 was manufactured commercially. TVA made a homogenous product by neutralization of nitric and sulphuric acids with ammonia. Advantages are less hygroscopicity, a good N:S ratio, and a combination of ammoniacal (75 per cent) and nitrate (25 per cent) N suitable for both spring and fall applications.

Equimolecular mixtures of ammonium sulphate and ammonium nitrate analysing 26-0-0-12 are produced on a significant scale in Western Europe.

Urea-ammonium sulphate, a physical mixture of these two materials, may be made by a variety of methods. Grades ranging from 40-0-0-4 to 30-0-0-13 have been made by oil prilling ammonium sulphate fines contained in molten urea by TVA. Handling and storage characteristics are said to be superior to urea alone. TVA has also done pilot granulation tests of 40-0-0-4 by granulating by-product ammonium sulphate with urea melt. A 34-0-0-11 grade made as a mechanical mixture or as a homogenous product by granulating AS fines with molten urea, has been sold. The mechanical method is preferred on economic grounds. Handling and storage are slight problems but not sufficient to warrant the expense of coating.

Urea sulphate and urea-gypsum are two of several co-ordination complexes which can be prepared from urea and sulphate salts. They doubtless form *in situ* in many complex fertilizers incorporating urea as well. There is no knowledge of their commercial availability, or of recent work done with them.

K-S AND K-Mg-S MATERIALS

Potassium sulphate (0-0-50-17) and potassium-magnesium sulphate (0-0-22-22) (11.6 per cent Mg) are important sources of plant nutrient S. They have special uses, e.g., when chloride-free K is preferred, as for tobacco, or in areas where the chloride-ion is unwanted because of prevailing high levels of soluble salts in the soil. They also supply K, S, or Mg individually or remedy multiple deficiencies of these elements. They are adaptable to direct application, bulk blending and suspensions.

N-P-S MATERIALS

Ammonium phosphate-sulphate (16-20-0-14) is the most popular grade. It contains about 40 per cent MAP and 60 per cent AS. Several processes for manufacture exist, e.g., the reaction of a mixture of phosphoric and sulphuric acids with ammonia; and the introduction of ammonium sulphate solutions and sulphuric acid into a phosphoric acid plant reaction circuit. A similar grade offered is 13-39-0-20.

A granular product (15-30-0-8) is currently manufactured, based upon ammonium phosphate, ammonium sulphate and gypsum.

Experimental quantities of a homogenous 23-23-0-6 product, based upon ammonium phosphate, ammonium sulphate and urea have also been made.

N-P-K-S MATERIALS

Any complete fertilizer making use of ammonium, potassium, or potassium-magnesium sulphate is automatically an NPK plus sulphate-sulphur material. Many manufacturers are exploiting this sometimes newly discovered aspect of their products.

OTHER SULPHATE SALTS

Gypsum is used as a source of S in the area where it is mined or produced as a by-product. Its low nutrient analysis, 13 to 19 per cent S, limits the extent to which it can be profitably used for fertilizer purposes. Magnesium sulphate, 20 per cent Mg and 27 per cent S, is occasionally used to supply Mg in clear liquids and foliar sprays. Several micronutrients, as sulphate salts, supply incidental S in a plant-available form wherever they are used. Lastly, there is sulphuric acid, but, like gypsum, it is primarily used as a soil amendment.

Problems related to mining, crystallization, etc., of sulphate materials are beyond the scope of this paper. Problems in making fertilizers from them are related to segregation and caking.

SULPHUR IN OTHER FORMS

Ammonium thiosulphate

Ammonium thiosulphate solution, 12-0-0-26, is the S product most widely used in clear liquid fertilizers today. It is prepared by reaction of SO_2 and aqueous ammonia followed by further reaction with elemental S, although variations on this basic method exist.

Solutions of ammonium thiosulphate are compatible in any proportion with neutral to slightly acidic ortho- and polyphosphates, the aqueous ammonia and with nitrogen solutions. It is not used with anhydrous ammonia. Blending of a wide variety of NS, NPS and NPKS formulations containing up to 10 to 12 per cent S is usually feasible, although trial blends are recommended, especially for potash grades, before going to large-scale mixing. Storage temperatures of ammonium thiosulphate are between —1 and 38°C. Steel and aluminium are suitable materials for storage containers.

Ammonium thiosulphate is applied directly to open ditch and sprinkler irrigation systems as well as to row crops. It has favourable soil amendment properties too, but there are less costly alternatives.

AMMONIUM POLYSULPHIDE

Ammonium polysulphide, 20-0-0-45, is a highly alkaline (pH 10) reddish-brown solution with a strong odour of H_2S. It is made by reacting aqua ammonia with H_2S and elemental S in the presence of heat and agitation.

The product is compatible with anhydrous ammonia but not with acidic solutions in which it decomposes to yield H_2S and colloidal S. Generally, it is not considered suitable for mixing with phosphate solutions although there are exceptions. Complete liquids containing up to 7 per cent K_2O have been made. Black iron, stainless steel, or aluminium equipment should be used for handling. Storage temperatures should be between —1 and 38°C. While competing with ammonium thiosulphate in many plant nutrient S applications, ammonium polysulphide is often used primarily as a soil amendment. It is applied in open ditch irrigation systems, injected into the soil, and when diluted, applied directly to the soil by spray applicators.

Urea-ammonium polysulphide solution, 25-0-0-35, is marketed and applied in irrigation water to vegetable crops on problem soils. The reasoning for its formulation is the feeling that some N is beneficial, in addition to the soil amendment properties of the polysulphide.

SULPHUR DIOXIDE

Liquid SO_2 (50 per cent S) is a potential high analysis fertilizer material. It is a pressure liquid (26 psi at 16°C), requiring special application equipment similar to that for anhydrous ammonia. Favourable agronomic results have been reported, but this material has not realized its potential, largely because it is incompatible with anyhdrous ammonia. It needs a separate tank and pump on application rigs, and the small amounts required for most cropping situations are difficult to meter.

Liquid SO_2, however, is becoming widely used indirectly, by generating the gas in portable S-burning equipment, and injecting it into irrigation waters to improve soil permeability and supply incidental plant nutrient sulphur. It is also used to reclaim tile drains clogged with Fe and Mn compounds.

Ammonium bisulphite (8.5-0-0-17) is available commercially as a 55 to 60 per cent solution. It is a by-product of smelting operations, made by absorbing SO_2 in aqueous ammonia. It is a non-pressure solution with a

TABLE 6-10
Recommendations for use of solid fertilizers containing sulphur

Material	Recommended use	Remarks
Ammonium phosphate-sulphur Ammonium polyphosphate-sulphur Ammonium phosphate-urea phosphate-sulphur Concentrated superphosphate-sulphur urea-sulphur	For direct application and bulk blends apply materials several months before beginning of growing season.	If used in starter fertilizer or shortly before beginning of growing season, some readily available sulphate should be included (15 to 20 per cent of the total sulphur applied).
Granular sulphur assemblages Flake sulphur	For direct application and bulk blends apply materials several months before beginning of the growing season. Fall applications should be encouraged.	Where feasible incorporate into soil four or five months prior to planting. When applied just in advance of planting or on severely sulphur deficient soils, some readily available sulphate should be included—see above.
Ammonium sulphate	For direct application and to some extent for bulk blending. Should be effective at almost any time.	Tends to segregate in bulk blends unless physical properties are improved by granulation. Where significant leaching losses are expected, apply shortly before planting or beginning of the growing season.
Ammonium nitrate-sulphate Ammonium phosphate-sulphate Ammonium phosphate-sulphate gypsum Ordinary superphosphate Potassium sulphate Potassium-magnesium sulphate	For direct application and bulk blends. Should be effective at almost any time.	Where significant leaching losses of sulphate are expected, apply shortly before planting or the beginning of the growing season.
Calcium sulphate (gypsum)	For direct application. Should be effective at almost any time.	Difficulties may be encountered in application (dustiness, clogging).

TABLE 6-11
Recommendations for use of fluid fertilizers containing sulphur

Material	Recommended use	Remarks
Ammonium thiosulphate	For direct application and blending with most fluid fertilizer products. Can be broadcast prior to planting or applied in starter fertilizers. Can be topdressed on certain growing crops. It can also be added through open ditch and sprinkler irrigation systems. Should be effective almost any time.	Can be blended with all neutral fluid phosphate products now available, all nitrogen solutions (except anhydrous ammonia) and most micronutrient solutions.
Ammonium bisulphite	For direct application and blending with most other nitrogen and nitrogen-phosphate fluids. Can be sprayed or dribbled on prior to planting. Can also be added through open ditch and sprinkler irrigation systems. Should be effective at almost any time.	Can be blended with all neutral fluid phosphate products now available and all nitrogen solutions except anhydrous ammonia. In many instances it is applied simultaneously with anhydrous ammonia, the two products from different tanks are taken to the ground through separate metering systems and injected into soil on common applicator knives.
Ammonium polysulphide	For direct application and blending with other nitrogen solutions. It is frequently injected into soil. Broadcast spray applications are possible following dilution with water. Single preplant applications are effective. Repeated applications at low rates are often made to growing crops through open ditch irrigation systems. Should be effective at almost any time.	Ammonium polysulphide is generally not considered suitable for mixing with phosphate-containing fluids.
Sulphuric acid	For mixing with wet process ammonium polyphosphate and anhydrous ammonia in the preparation of clear liquid blends. Should be effective at almost any time.	Sulphuric acid has been applied directly to crops such as onions and garlic for weed control purposes.
Suspensions containing elemental sulphur	For direct application and for simultaneous application with other fertilizers, the suspensions should be applied several months before beginning of the growing season.	If used in starter fertilizer or shortly before beginning of growing season, readily available sulphate should be included (15 to 20 per cent of the total sulphur applied).
Suspensions containing one or more sulphate salts such as ammonium sulphate, potassium sulphate, potassium-magnesium sulphate and ammonium phosphate-sulphate	Should be effective at almost any time.	Where significant leaching losses are expected, apply shortly before planting or the beginning of the growing season.

salting-out temperature of about 0°C. In combination with aqueous ammonia or N solutions any N to S ratio may be obtained. A number of NPS grades may be prepared with nonacidic ammonium ortho- and polyphosphate solutions.

Mild steel, aluminium and plastic vessels are suitable for storing ammonium bisulphite. When mixed with other solutions, such as aqueous ammonia, an inhibitor is used to prevent corrosion in mild steel containers.

ULTRA HIGH ANALYSIS MATERIALS

Four groups of compounds, several of which can be formulated to contain S, are being evaluated at TVA. These are: vapour-phase reaction products, linear polyamides, phosphonitrilics and cyclic metaphosphimates. The first group includes reactions of ammonia with elemental P and S, and with phosphorus pentasulphide. Analyses of 7 to 28 per cent N, 60 to 97 per cent P_2O_5 and 12 to 56 per cent S showed promising availability to crops.

Of the amides, one monomer, 38-64-0-29, was an excellent fertilizer material. Some polyamides exhibited slight toxicities at high application rates believed to be caused by H_2S evolved from hydrolysis. Sodium and ammonium derivatives of the monomer $[SP(NH_2)_3]$ were effective sources of N and P in greenhouse tests, although their S effectiveness was not evaluated.

One of the most promising ultra high analysis compounds is phosphonitrilic hexaamide. A related compound of the general formula $(NH_2)^4NHP_2N_2SO_2$ containing 14 per cent S has exhibited no toxicity and possesses acceptable agronomic properties.

Although these and like materials may not be commercially available for some time, if ever, fertilizer technologists should be aware of these developments which may affect them in the future.

Summarized recommendations for use of the more important sulphur fertilizers, drawn from Beaton and Fox (2) and the Sulphur Institute (3), are shown in Tables 6-10 and 6-11. They contain technological as well as agronomic aspects relating to their use.

REFERENCES

(1) *Proceedings*, 12th and 13th Technical Conferences of N.Z. Fertilizer Manufacturers Association (1969, 1970).

(2) Beaton, James D. and Fox, Robert L. (1971). 'Production, marketing, and use of sulphur products' in Olson, R. A., Army, T. J., Hanway, J. J. and Kilmer, V. J. (eds), *Fertilizer Technology and Use*, 2nd edn, pp. 335-79. Soil Science Society of America, Madison, Wisconsin. 611 pp.

(3) Bixby, D. W. and Beaton, J. D. (1970). 'Sulphur-containing fertilizers'. Tech. Bulletin No. 17, The Sulphur Institute, Washington and London. 27 pp.

Prescription of the Types of Sulphur Fertilizers Currently Needed in Agriculture

A. JOHNSON

There are various factors which must be taken into account when attempting to prescribe the types of sulphur fertilizers currently needed in agriculture. These include such things as nature of the soil, type and purpose of crop or pasture, method and frequency of application, climate, materials available, and economy and convenience as far as the user is concerned. All these may be summarized as *efficiency* for the specific conditions of use; *economy* in terms of costs of material, transport and application; and *convenience* in terms of fitting in with normal fertilizer operations.

Taking the last of these first, in the vast majority of situations, plant nutrient S as an ingredient in regular fertilizer treatment is most likely to meet the requirement. In relatively few situations will the farmer of the future be interested in straight S fertilizers.

As far as economy is concerned, it is only to be expected that the user will be looking to get his S, as well as other nutrients, on the ground at least possible cost to him. At the same time, all those involved in the various parts of the operation, through from the supplier of raw materials to the spreader of the finished product, need to be adequately rewarded for capital and labour involvement. In the past, S has been regarded largely as a 'give away' plant nutrient that has been included in quite a few fertilizers in common use, its presence in the product being promoted or regarded as a bonus. With S given its correct place as an important and often deficient plant nutrient it will rightfully be priced as such and thus share its part of the cost burden. In this event, the user then has every right to prescribe what his requirements are and expect the prescription to be met.

However, because of the large number of factors involved, some of which have been listed earlier, it must be realized by all concerned, user as well, that there is little possibility of supplying all plant nutrients, including S, as fertilizer in an ideally balanced combination for each specific situation. Apart from any other consideration, for a great many situations this ideal combination will not be known.

This should not prevent us from attempting to define a number of commonly occurring situations where the need for S is specified along with one or more of nitrogen, phosphorus and potassium and a range of fertilizers developed to meet the varying needs revealed.

EFFICIENCY CONSIDERATIONS

For any S fertilizer to do the job required of it efficiently it must first meet certain agronomic requirements, using the term 'agronomic' in its widest sense. These have been covered in detail in previous sessions of this symposium and are dealt with very briefly in this paper as a back-drop to more detailed consideration of specific fertilizers.

Sulphur is used by plants in the sulphate form

For applied S to be taken up by plants it must be in the sulphate form at the time of application or else oxidized to this form. If applied in the sulphate form this is an advantage to the plant at the time of application but the plant may later be at a disadvantage because sulphate S is readily leached. An additional disadvantage from the economy point of view is that the S is not in a concentrated form. Elemental S is the most concentrated S carrier. From the practical angle, elemental S of particle size smaller than that passing through a 60 mesh screen can be just as effective as the sulphate form in terms of crop response in the year of application.

Sulphur requirement is influenced by high application rates of other nutrients

S deficiencies become more numerous and more limiting to yield performance as rates of application of N, P and K are increased. Fertilizer S requirements in intensive agricultural situations can be determined on the basis of the amount of N and P applied as fertilizer. On soils which are both N and S deficient 1 kg S for every 5 to 7 kg N will be satisfactory. Where P and S are deficient the relative amounts will be 1 kg and 0.75 kg. These relationships apply only when the amounts of N and P meet the crop's normal requirements. If low rates are being used the amount of S included is likely to be inadequate.

Crops differ in their requirements for S

The relative requirements of different agricultural plants have been adequately covered in a previous session but will obviously have a major bearing on the matter being dealt with.

Crop S requirements increase as higher yields are obtained

This is adequately explained by the Law of the Minimum.

The purpose for which a crop is used affects S requirement

Annual cash crops require some of their S in a readily available form with the remainder becoming available fairly quickly, whereas pastures and forage crops require some S in a readily available form at the commencement of their growth cycle with the remainder becoming available continuously over a longer period of time.

Protein content is affected by early availability of S

The yield of protein in forage and hay crops as well as pastures is affected at an earlier stage in plant growth than yield of dry matter. This emphasizes the need of readily available S early in the growth cycle of such crops.

Leaching of S is affected by soil type and rainfall

Sulphate S is readily leached from the surface horizons of many soils, especially those of coarse texture in high rainfall areas.

Method of application may have a bearing on form of S used

Surface broadcasting of the sulphate form will not usually reduce its effectiveness. Mixing reduced forms with soil will generally result in most effective use, especially in the case of annual crops.

A RANGE OF S FERTILIZERS REQUIRED

Just as a range of fertilizers containing various amounts and different forms of N, P and K are needed to meet the varying conditions and requirements of Australasian agriculture, so there is a need for a range of S-containing fertilizers in which both the form and amount of S are varied.

As greater emphasis is placed upon production per hectare and per unit of labour it is essential that greater attention be given to the increasingly important role of the S in fertilizers. An examination of the fertilizers currently available gives the impression that many of these are manufactured because they happen to fit in well with the producers' specifications and plant rather than the users' requirements. This applies particularly as far as S is concerned. The time has now certainly arrived when much more attention must be given to the efficiency of fertilizers in supplying S in the required form or forms together with economy and convenience in their use. In the majority of cases this will mean modification of manufacturing processes or further treatment of currently manufactured products. In this way a range of products will be made available which better meets the needs of current agriculture at a price which recognizes the increasingly important place of S in crop and pasture production.

S REQUIREMENTS IN FERTILIZERS FOR SPECIFIC PURPOSES

Fertilizers can be conveniently classified into three main categories according to time of use in relation to crop or pasture growth cycle. These are pre-sowing, starter and post-sowing.

Pre-sowing fertilizers

Fertilizers falling into this category may be applied from several months to a day or so before sowing. The main ones involved in Australasian agriculture are anhydrous ammonia and urea, both high analysis nitrogenous products for use primarily in intensive systems of agriculture, particularly where irrigation is available.

Anhydrous ammonia is normally applied at some depth several months to several weeks before sowing and has a high convenience factor in concentration and being able to be applied in slack periods, often in conjunction with a tillage operation. In known S deficient situations and where high yields are aimed at, which would normally be the case, this fertilizer could be greatly improved by the inclusion of elemental S. A 74-0-0-10 product has substantially the same physical properties as anhydrous ammonia. Injection under pressure causes the S to be deposited in a very finely divided form which will be oxidized over a period of time. Where injection of this product takes place some months prior to sowing a high analysis starter fertilizer with little or no S may be used. However, where placement is very deep or injection carried out shortly before sowing, use of this product should be followed by a starter fertilizer which has some S in a readily available form for use by the crop during the establishment stage.

Urea-sulphur, applied and mixed with the soil in the same kind of situations as anhydrous ammonia, would give this product an advantage over straight urea. For long season crops or where the urea-sulphur is applied some weeks before sowing a 40-0-0-10 product with added S in the elemental form would be suitable. For short season crops or for application just prior to sowing, a 35-0-0-10 product, using ammonium sulphate as the source of S, would be appropriate allowing use of a low S starter or one in which the S is present in the elemental form.

Where high value per hectare perennial crops are being grown and a 'fertilize the soil' rather than a 'fertilize the crop' approach is taken, a mixture of finely ground phosphate rock and elemental S in either the powder or molten form may offer possibilities. Such mixtures worked into the soil well before sowing or planting could result in the phosphate rock becoming more readily available over a period of time as a result of sulphuric acid formation, as well as providing a source of S.

Starter fertilizers

These are normally compound fertilizers containing N and P with or without K. However, under Australasian conditions, various grades of superphosphate come into this category also.

With several exceptions the starter fertilizers available are satisfactory in terms of the amount of contained S. The exceptions are double and triple superphosphates, ammonium phosphates and a product having low N, medium P and high K (6-12-23-0). Unfortunately, whilst low in number these, particularly the first three, are widely used as starter fertilizers especially for cereal crops.

The various grades of superphosphate are covered in some detail in the following section dealing with post-sowing fertilizers. It is sufficient to indicate at this stage that the S content (12 per cent) of single superphosphate is satisfactory, though all the S is in the sulphate form. In double superphosphate it is 4.5 per cent S and in triple superphosphate it is 2 per cent S, in the sulphate form in both cases. It is considered that a superior product, replacing single, double and triple superphosphates as sowing fertilizers in a great many situations, would be 0-20-0-15, with 15 per cent of the S (elemental) in particle sizes finer than 80 mesh and the remainder of the S in varying particle sizes but with emphasis on those finer than 40 mesh. In situations where P is deficient and the soil N level satisfactory or brought to a satisfactory level by pre-sowing or post-sowing application of a nitrogenous fertilizer, such a product would have wide use in Australasian agriculture, particularly in cereal and legume-based pasture areas which account for a very high proportion of the area where fertilizers are used.

The ammonium phosphates such as 12-22-0-3 contain S in relation to P in such low ratios that the S content can be virtually discounted. These would be greatly improved as starter fertilizers where a small amount of N and S is required if they contained up to 10 per cent S of fine particle size.

The continued use of straight anhydrous ammonia or urea in combination with ammonium phosphates or triple superphosphate, as occurs with many irrigated crops such as maize, grain sorghum and cotton, must eventually bring about a situation of S deficiency which will be reflected in lower yields and/or products of lower protein content.

The 6-12-23-0 product referred to earlier was designed as a starter fertilizer for pastures in K deficient areas. Whilst there is undoubtedly a place for a high K starter fertilizer of this type in K deficient areas, such as parts of the south coast of New South Wales, it would be greatly improved by incorporating S of fine particle size to the extent of 8 to 10 per cent or using potassium sulphate as the source of K.

As stated earlier, most of the remaining starter compounds or mixtures are satisfactory as far as their S content is concerned though all of this is in the sulphate form and improvement could possibly be effected by fortification or replacement to some degree with elemental S.

Post-sowing fertilizers

These are mainly fertilizers which contain either N, P or K though, in some instances particularly in the case of perennial crops such as fruit trees and vines, they may contain all three of the above plant nutrients. Any S present is usually quite incidental and a result of the manufacturing process used. As far as the compounds are concerned, even though the S component is incidental it is usually present in adequate amounts in the sulphate form. As two or more applications are frequently made in the annual fertilizer programme, the S situation is probably adequately catered for.

It is with the fertilizers that supply one of N, P and K, particularly the first two, that considerable improvement as far as S is concerned is warranted.

The three most commonly used straight N fertilizers for post-sowing application or top-dressing are ammonium sulphate, ammonium nitrate and urea. The first of these has an excessive amount of S in relation to N and all this in the sulphate form. The latter two have no S at all.

A mixture of ammonium sulphate and ammonium nitrate would yield a 27-0-0-12 product. Use of this would reduce the wastage of S which almost certainly occurs in most applications of ammonium sulphate and still maintain at least part of the desirable features of ammonium nitrate. Such a product would be of considerable value for top-dressing grassy pastures on high P soils as well as forage, hay and silage crops.

For those situations requiring a greater concentration of N and lower S, replacing a quarter of the ammonium nitrate with ammonium sulphate would yield a product of approximately 30-0-0-6 analysis, retaining still more of the desirable features of ammonium nitrate and supplying S in a ratio with N required in many situations for good yield and protein content. This product would have a particular place for top-dressing or side-dressing cash crops such as sorghum, maize and cereals where application of a nitrogenous fertilizer has been deferred due to unfavourable conditions at sowing or where a N deficiency becomes apparent after a crop has been established.

Urea combined with ammonium sulphate to give 35-0-0-10 provides a product which could well enhance the value of urea for top-dressing of pastures, forage and cash crops where irrigation is available, or in side-dressing operations where the fertilizer is mixed with the soil.

Single superphosphate has been with us as long as inorganic fertilizers have been used. It has played a very prominent part in Australian agriculture based as it is and has been on legume-based pastures in predominantly N, P and S deficient situations. It has also, unfortunately, placed a great many people in the 'hundredweight or bag (whatever size that may be) per acre' mental straight jacket.

With the introduction of the metric and elemental systems in the fertilizer industry, single superphosphate should be laid to rest with the old systems. This would help to remove the mental straight jacket and place fertilizer use on a more rational basis as well as making way for a more suitable product.

There is certainly a place for several grades of superphosphate, but based on a high analysis starting point and with variation of P and S contents around this point.

Single superphosphate (0-9.5-0-12) requires replacement with a product such as 0-20-0-15, with 15 to 20 per cent of the S in the sulphate or fine-sized particle elemental form and the remainder in the elemental form of varying particle size but with emphasis on finer than 40 mesh. This would be an excellent pasture top-dressing fertilizer over a very wide area for soils of low P and low to medium S status. It would also be an excellent sowing fertilizer to replace single, double and triple superphosphates as stated earlier.

For soils of high P and medium to low S status a sulphur-fortified product such as 0-10-0-35 would meet the situation, providing attention is paid to S particle size with 20 per cent finer than 80 mesh. This would replace SF 25 superphosphate (0-7.7-0-27), previously known as SF 30, in the pasture top-dressing programme. It can be anticipated that many pasture areas which have been top-dressed regularly with superphosphate or its suggested replacement will move into a satisfactory soil P situation where biennial applications of 0-10-0-35 at 150 kg per ha or less (not 1 cwt per acre) will provide all the P and S required in a satisfactory manner.

As single superphosphate would no longer be manufactured, for very high P and very low S status soils it would be possible to replace SF 45 (0-6-0-46), previously known as SF 50, with elemental S added to gypsum in the same ratio as elemental S is now added to single superphosphate to produce SF 45. This product, 0-0-0-45, or sulphur-fortified gypsum would retain some of the S in the sulphate form and the remainder would be in the elemental form in particles of varying size. This would be a very suitable pasture top-dressing product for certain areas in the Monaro and Upper Hunter districts of New South Wales. From the economic viewpoint, straight gypsum as a direct application material for correcting S deficiency should be banished with single superphosphate.

Having successfully disposed of several straight nitrogenous fertilizers as well as single, double and triple superphosphates, attention can now be directed to the straight potassium fertilizers. In this section, as with the nitrogenous fertilizers, there is either a feast or a famine. Potassium sulphate is

0-0-40-16, with all the S in sulphate form. Potassium chloride is 0-0-50-0. Potassium sulphate could be used should there be any suggestion of S deficiency, which has not already been corrected, accompanying the K deficiency. By crediting this fertilizer with the value of its S content it becomes a more attractive proposition price-wise in situations where the contained S is required.

OTHER FORMS AND USES OF SULPHUR IN FERTILIZERS

Emphasis in this paper has been placed upon the sulphate and elemental forms of S. It is believed that these two forms have the greatest general use in agriculture because the sulphate form is the one that is most readily available for plant use and the elemental form is the most concentrated and at the same time, with proper regard to particle size, can be used to meet S deficiency in a variety of situations. This does not necessarily mean that all other forms of S should be disregarded and left out of calculations, particularly if they are effective and in plentiful local supply.

No mention has been made in this paper of liquid fertilizers containing S except for anhydrous ammonia-sulphur. Whilst there is no doubt that there will be increasing use of liquid fertilizers in Australasian agriculture this is looking somewhat in the future and this paper is primarily concerned with current needs.

There is one aspect, however, where the plant nutrient value of S could be exploited, at the same time making use of it to control release of other nutrients. S-coated urea is one such possible product and a further possibility is S-coated potassium chloride. In this latter case the advantages could well be reduction of luxury consumption of K, reduced leaching losses of K on sandy soils, some avoidance of high concentrations of chloride as well as providing S for plant use.

Various granular S assemblages are possible though little use is seen for these as direct application materials except perhaps to replace sulphur-fortified gypsum. Should bulk blending become more common in fertilizer formulation in this part of the world, however, these assemblages could play an important part and would perhaps help to overcome technological problems which may arise in manufacturing some of the products suggested.

The principle behind these assemblages is that they contain additives which cause disintegration at a fairly rapid rate after they are applied to the soil, releasing S of small particle size to ensure rapid oxidation. Whilst the quantities of the various additives may vary widely it is desirable to keep the S content as high as possible whilst still retaining good disintegration characteristics.

Acknowledgement. In preparing this paper publications of the Sulphur Institute, Washington, U.S.A., have been freely drawn upon. This is gratefully acknowledged.

7 Conclusion

Summary Review

S. L. TISDALE

In summarizing this Conference I should firstly like to review briefly the papers which have been presented; secondly, what appears to me to have been the main accomplishments of this meeting; and thirdly, to set forth some of the questions which have come to my mind as a result of papers which have been presented.

The stage for the entire meeting was appropriately set by Dr Lloyd Evans' paper. It began with an historical review of the use of sulphur down through the ages, in agricultural and other areas, and ended with an overall statement concerning the importance of sulphur in the agriculture of both Australia and New Zealand. It was itself an excellent summary of the entire Conference. This was followed by Dr Probert's paper on the occurrence of sulphur deficiency in Queensland and the Australian tropics. While the need for sulphur has not been as widely demonstrated in these as in some other areas of Australia, there are none-the-less numerous instances of sulphur deficiency in important areas of the tropics. These deficiencies have been largely observed on pasture species, particularly tropical legumes. In some cases, lucerne has also been found to respond to sulphur application.

Dr Blair then discussed the occurrence of sulphur deficiency in the remainder of Australia. He pointed out those areas in which deficiencies have been observed, presented a summary of the magnitude of the responses, and attempted to derive some pattern in their occurrence. Professor Walker next discussed the occurrence of sulphur deficiencies in New Zealand, and stated that had it not been for the long continued use of normal superphosphate (containing 10 to 12 per cent sulphur) deficiencies in this element would be of almost universal occurrence in both the North and South Islands of New Zealand.

Only comparatively recently has the need for sulphur by forest species been demonstrated. Mr Humphreys summarized the results of his work with *Pinus radiata* in Australia. This work, in my opinion, is of significance not only to the production of forest species in Australia but also in New Zealand where timber production contributes significantly to the total agricultural output and where as has already been pointed out sulphur deficiencies are quite widespread.

Sulphur and its compounds have long been used in many areas of the world for the reclamation of alkali and saline-alkali soils and for the treatment of water used to irrigate such areas. Dr Loveday discussed the use of gypsum as a soil amendment in the irrigated alkaline-calcareous areas of Australia and suggested the need for additional research on this topic. I agree wholeheartedly with Dr Loveday and would like to suggest that in addition to gypsum, the use of other sulphur compounds could well be investigated. Gypsum is an excellent soil corrective, but it has the disadvantage of having to be used in unusually large quantities in many cases, contributing to the expense of the reclamation operation. Many higher analysis sulphur compounds, for example, ammonium polysulphide, elemental sulphur and even SO_2 have been shown to be equally as effective as gypsum when applied at rates equivalent to the sulphur contained in the gypsum.

The next session was devoted to a discussion of three rather basic topics relating to the biogeochemistry of sulphur, the nature of sulphur in soils and the transformations of this element in soils. These three topics were very well covered by Drs Trudinger, Williams and Freney and the inclusion of these three papers in the Proceedings will afford a ready reference for those interested in the basic geological and soil chemistry of this element. It was followed by Barrow, Weir and McLachlan's papers dealing with the reaction of fertilizer sulphate in soils, the oxidation of elemental sulphur and sulphides in soils, and the residual effects of various sulphur fertilizers respectively. These papers in my opinion are of considerable importance to agricultural advisory officers and to people in the fertilizer industry because the information they contain makes possible the more intelligent and efficient use of the various types of sulphur-containing fertilizers.

The programme then turned to a discussion of sulphur cycling in soils under grazed pastures by Mr Till of the Animal Physiology Division in Armidale. This paper, and the previous ones dealing with the transformations of organic sulphur in soils, raised far more questions than they answered and before the relationship between plant-available sulphur and that in the sulphur pool can be fully established, a considerable amount of additional research will obviously have to be carried out.

The importance of sulphur in the nutrition of ruminant animals was next discussed in two papers, one by R. J. Moir and the other by A. M. Downes. Dr Moir's work has shown quite clearly that the amount of nitrogen recycled and utilized by ruminant animals, particularly sheep, is greatly affected by the amount of sulphur in the diet. He has shown in fact that dietary N:S ratios of 9.5 to 10 are needed for the most effective utilization of nitrogen by sheep. When this fact is considered with Mr Downe's comment, to the effect that there had been no demonstrated response to supplemental feeding of sulphur to grazing sheep, the need for additional research in this area certainly suggests itself. When one reviews the wide range of N:S ratios in subclover and other pasture species, reported in the literature, it is difficult to understand why responses to the supplemental feeding of sulphur compounds to grazing sheep have not been observed.

The uptake and translocation of sulphur in plants, the function of sulphur in plant growth and metabolism, and the sulphur requirement of plants was treated in the next three papers. The fundamental chemistry and behaviour of sulphur in plants covered by Drs Bouma and Anderson will serve as excellent references on this topic. Mr Spencer's comments on the sulphur requirement of plants should prove helpful to the agricultural advisory services and fertilizer industry in the formulation of meaningful fertilizer recommendations for the grower and the grazier.

Three papers dealing with the use of plant and soil tests as diagnostic tools for the assessment of sulphur requirement were given by Drs M. B. Jones and H. M. Reisenauer of the University of California and Dr W. M. H. Saunders of New Zealand. There is apparently a considerable diversity of opinion as to the value of these two types of tests in determining the need for sulphur fertilization and it is obvious that additional work is needed to determine more clearly the conditions under which each of these types of tests can be successfully employed. Perhaps the use of both of these diagnostic techniques in conjunction with a crop logging system should be considered. In my opinion, it is unlikely that in the foreseeable future any one or two tests can be developed which will be used universally under all soil conditions and for all crops. As pointed out by Dr Reisenauer, the fact that levels of soil sulphur can be predicted by the use of soil tests a relatively high proportion of the time suggests that these tests, while not perfect, are far better than nothing and that their use should not be discontinued, but should be studied more intensively with the hope of even further improvement in their predictability.

The six papers presented in the latter part of this session provided an excellent overview of the methods of soil and plant analysis which are being used in the various states of Australia and has given us a good insight as to the opinions held concerning the worth of these tests. The vast amount of work which has been done has been effectively summarized in the Handbook, edited by Mr McLachlan and distributed in association with this Symposium. In my opinion, this is one of the most useful publications I have seen relating to the need for sulphur fertilization in an area, and I believe that agricultural advisory officers and people in the fertilizer industry especially, will find this book of considerable value to them in their work. It contains much previously unpublished data on the deficient areas in the various states of Australia and the types of crop responses which have been observed. This will enable individuals in the various states to understand more clearly what their fellow workers have been doing, and what they themselves need to undertake to answer the problems which may be arising in their respective areas. I gather from the comments and discussion concerning these papers that there is a considerable difference of opinion as to the value of the tests themselves; a rather wide divergence among the types of tests which are being used; and quite a difference in the extent to which the need for sulphur in the various states has been investigated. I am sure that the papers included in this book, along with the opinions that have been expressed by the various authors, will serve to generate additional research which hopefully will answer some of the many questions that seemed to arise during this session.

Finally, a series of interesting papers dealing with the fertilizer materials available and those being used in Australia and New Zealand was presented by Mr Bell and Dr Rogers. Most of the fertilizers produced and used in each of these countries contain adequate sulphur, although there is some rather limited production of materials such as triple superphosphate and ammonium phosphate. Nitrogen fertilizers which are used in Australia, and to a lesser extent in New Zealand, are anhydrous ammonia, urea and ammonium nitrate.

None contains sulphur. Large amounts of ammonium sulphate are available, particularly in Australia, and this material is an excellent source of plant nutrient sulphur.

Dr Swaby presented the results of the work which he has been doing for several years on material known as 'biosuper'. This product appears to have real promise in hot humid countries, where the temperature and rainfall are sufficiently high to permit the sulphur oxidizing organisms, with which the material is inoculated, to function properly. In my opinion, it is doubtful whether 'biosuper' will ever find a place in countries which have a sophisticated acid-based fertilizer industry. However, in developing areas of the world, particularly those in the hot humid regions, 'biosuper' may find a place in the development of their agriculture. I do not wish to get involved in the pros and cons of the commercial development of this product. This will be taken care of by the economics which dictate its use. However, I do think Dr Swaby's work is significant for among other reasons, 'biosuper' is by far the best of the many rock phosphate-sulphur assemblages that have been developed in past years.

New fertilizer materials which have been developed, some of which are in commercial use in the United States, were next considered by Dr Kilmer. He discussed those developed by TVA and others for use in both solid and liquid form. In addition to these, two others were discussed. They were solutions consisting of anhydrous ammonia and elemental sulphur and sulphur dioxide itself. In the latter case, sulphur dioxide is injected directly into irrigation water in the saline-alkali regions of the western United States. Its purpose is to increase the rate at which water moves into these irrigated soils.

The last paper given by Mr Johnson was an unusually interesting one, because it set forth what one large purchaser of fertilizer considered to be products that could be used profitably in his farming operation. Mr Johnson is unusually qualified to make such a presentation, as he has probably been on more sides of the agricultural operation in Australia than most of us here in this room. He has therefore seen all sides of the problems discussed here and has made some rather revolutionary suggestions which will set quite a few of us thinking. While I do not wish to project myself into the middle of any controversy, I think his comments will generate considerable thought, especially on the part of the fertilizer manufacturers in Australia and New Zealand regarding possible new sulphur-containing fertilizers which are better tailored to meet the needs of the wide range of soil and cropping conditions found in these two countries.

What has the Conference accomplished? Briefly it has:

1. Summarized existing information on the state of agricultural research on sulphur in Australia and New Zealand as well as the existing data on crop responses and sulphur deficient areas in both countries. The conference results suggest that while much information is available on these several topics there are many geographical areas in which additional and intensive research programmes are needed.
2. Brought about a face-to-face exchange of ideas among research workers, advisory officers and members of the fertilizer industry. This confrontation has stimulated thought on those problems needing additional study

and I would like to suggest that these confrontations may have resulted in a mild antagonistic symbiosis which will ultimately result in a better understanding of the various problems and lead to a quicker solution of them.

3. Resulted in the publication of two very valuable sets of Proceedings which can serve as references for research and advisory officers and for those people engaged in the production and sale of fertilizers.
4. Given an idea of the fertilizer materials that are needed and has suggested additional ones that could be investigated for possible introduction into both Australia and New Zealand.

The final part of this summary is concerned with questions which have been raised in my mind, and where work to date has not provided satisfactory answers. Some of these areas include:

1. Additional work on the mineralization and immobilization of organic soil sulphur, the so-called sulphur pool, and the impact that these forms have on the plant-available sulphur in the soil solution.
2. Studies on the effect of sulphur compounds other than gypsum on soil reclamation, the movement of water into soil and their effect on phosphorus and micro-nutrient availability in calcareous soils of both Australia and New Zealand, especially when these materials are applied in a band. The effect of banding applications of several of these acid-forming sulphur materials on the availability of soil phosphorus especially needs study, because of the possible reduction in the supply and use of phosphate fertilizers in the future. An investigation of the effectiveness of fertilizers with different N:S and P:S ratios under various soil and cropping conditions in both countries. Additionally an evaluation of the effectiveness of sources of sulphur other than normal superphosphate and elemental sulphur should be undertaken on a wider scale.
3. An economic evaluation of the effect of various fertilizer treatments on wool and meat production. Not one single report in this session dealt with the economics of applying phosphorus and sulphur at different levels to paddocks or field crops under different soil management and economic conditions. The acceptance of any farming practice by the grower or grazier is predicated upon its economic worth to him and I have seen little if any such research data from either New Zealand or Australia.
4. Volatile sulphur compounds are known to exist in plants. Studies to determine the effect that sulphur fertilization has on the production of these volatiles in pasture species needs to be undertaken, and whether or not these volatiles are related to palatability and hence intake by ruminant animals should be determined.
5. The greatest need lies in the area of co-operation between animal scientists and plant and soil scientists. We know that the sulphur content of plants can be increased by increasing the rate of applied sulphur, that this sulphur can accumulate in the tissue as SO_4, and that SO_4 in the rumen is utilized in the synthesis of bacterial protein. Moir and his co-workers have shown that increasing the sulphur in the diet relative to nitrogen can result in better nitrogen utilization by the animal, and others have shown that increases in the sulphur level in the diet can also result in greater cellulose digestion. It seems to me that what is needed now is to

determine whether an increased level of sulphur in pasture plants can improve wool and meat production, and for this much closer co-operation between plant and animal scientists is needed. This is not a unique requirement for Australia and New Zealand. It happens in most countries of the world, but it seems to me that the opportunity for such co-operation is unique in these two countries simply because of the high calibre of the research workers themselves and the proportion of the gross national income that is derived from agricultural pursuits, especially animal production.

6. The workers at the Animal Physiology Division at Prospect have demonstrated quite clearly that if high sulphur amino acids can be placed in the true stomach of sheep, wool production can be increased by 30 per cent or more. Limited success in effecting this has been obtained using several different techniques. None, I believe, is sufficiently attractive economically to entice a grazier with several thousand sheep to undertake any one of them. Even if a time is reached when sheep receive all of the sulphur they can possibly utilize in their diet, the amount of the sulphur containing amino acids reaching the true stomach will be limited by that which is synthesized by the rumen micro-organisms. So an impasse will have been reached even under such ideal conditions. To increase wool production even further, some economical way must be found to cross this barrier. So at the risk of engendering professional ridicule and the scorn of my colleagues, I would like to make a suggestion for possible future research in the area of animal biochemistry and microbiology. Is it feasible to attempt the development, through radiation or other techniques, of strains of rumen micro-organisms that are high producers of cystine and methionine? Radiation techniques have been used with some success in developing mutants in higher plant species, why not then in microflora and microfauna? If this can be done, infection of sheep with such mutants could be easily effected and from then on the job will have been done. Wild? Yes—undoubtedly, with many obstacles to overcome but being in the blissfully ignorant position of an agronomist, I advance the suggestion to my more learned colleagues in animal nutrition and microbiology.

The world shortage of petroleum may well reduce the production of synthetic fibres in the future. If this comes to pass, natural fibres will once again be in the forefront and it would be nice indeed to have flocks of sheep that could easily produce 20 to 30 per cent more wool than they do now.

Before concluding, I would like to acknowledge the outstanding job that has been done by Mr K. D. McLachlan and his colleagues in organizing and putting on this Symposium. Whether or not you agree with all that has been said here, I think all would agree on one thing, namely, that the organization and execution of this Symposium has been outstanding. I am sure that each of you would wish to join me in giving them a rising vote of thanks for their efforts on our behalf.

Index

INDEX